The Road to the Top on the Map: Conversations with Top Women of the Automotive Industry

The Road to the Top is Not on the Map: Conversations with Top Women of the Automotive Industry

Warrendale, Pennsylvania, USA

400 Commonwealth Drive
Warrendale, PA 15096-0001 USA
E-mail: CustomerService@sae.org
Phone: 877-606-7323 (inside USA and Canada)
724-776-4970 (outside USA)
FAX: 724-776-0790

Library of Congress Catalog Number 2019939936
http://dx.doi.org/10.4271/9780768000931

Information contained in this work has been obtained by SAE International from sources believed to be reliable. However, neither SAE International nor its authors guarantee the accuracy or completeness of any information published herein and neither SAE International nor its authors shall be responsible for any errors, omissions, or damages arising out of use of this information. This work is published with the understanding that SAE International and its authors are supplying information, but are not attempting to render engineering or other professional services. If such services are required, the assistance of an appropriate professional should be sought.

ISBN-Print 978-0-7680-0092-4
ISBN-MediaTech 978-0-7680-0093-1
ISBN-epub 978-0-7680-0095-5
ISBN-prc 978-0-7680-0094-8
ISBN-HTML 978-0-7680-0096-2

To purchase bulk quantities, please contact: SAE Customer Service

E-mail: CustomerService@sae.org
Phone: 877-606-7323 (inside USA and Canada)
724-776-4970 (outside USA)
Fax: 724-776-0790

Visit the SAE International Bookstore at books.sae.org

Chief Product Officer
Frank Menchaca

Publisher
Sherry Dickinson Nigam

**Director of Content Management/
Senior Production Manager**
Kelli Zilko

Production Associate
Erin Mendicino

Manufacturing Specialist
Adam Goebel

It's important not to limit your potential. Progress doesn't always travel in a straight line. Each new opportunity will broaden your skills and perspective, and the wider your experience, the more you bring to the table as you advance to leadership.

—Mary Barra
Chairman and CEO, General Motors

contents

CHAPTER 1

Sue Bai   1

CHAPTER 2

Susan Brennan   5

CHAPTER 3

Kim Brycz   11

Carla Bailo
President and CEO
Center for Automotive Research—Ann Arbor, MI

- Tel: 734-662-1287
- Email: cbailo@cargroup.org,
- https://www.linkedin.com/in/carla-bailo-604ba19/
- Twitter: @carlamobility
- Topics: automotive, smart mobility, smart cities

Education

- MS, University of Michigan, Mechanical Engineering, 1986
- BS, Kettering University, Mechanical Engineering, 1983

Carla Bailo is a leader in engineering and vehicle program management with 41 years of experience in the automotive industry. As the Center for Automotive Research (CAR) President and CEO, she leads the overall mission, manages day-to-day business, and is the face for the automotive industry in several venues. CAR is an independent, nonprofit research body respected for unbiased reporting on issues related to HAV's (highly automated vehicles), lightweighting and materials research, powertrain and electrification, and business/government economic impacts on the automotive industry.

In addition to her role at CAR, Carla serves on the National Academies Committee on Assessment of Technologies for Improving Fuel Economy of Light-duty Vehicles and is a board member for the National Academies BEES (Board on Energy and Environmental Systems)." Further, she is an independent director for SM Energy. She serves on the following nonprofit boards:

1. University of Michigan College of Mechanical Engineering External Advisory Board
2. The Ohio State Engineering College of Engineering External Advisory Board
3. Engineering Society of Detroit Board of Directors
4. Michigan Manufacturing Technology Center Board of Directors
5. Clemson University University Transportation Center Advisory Board

In her role prior to CAR, she was Ohio State's assistant vice president for mobility research and business development. In this role, she implemented the university's sustainable mobility and transportation innovation, while integrating related research and education across Ohio State's academic units. She also led Ohio State's involvement as the primary research partner for Smart Columbus, a $140 million program to transform Central Ohio into the nation's premier transportation innovation region.

Carla was the 2016-2018 vice president of Automotive for SAE International, a global association of more than 138,000 engineers and related technical experts in the aerospace, automotive, and commercial-vehicle industries.

Carla has 41 years experience in the automotive industry with 25 years at Nissan. In her most recent role at Nissan, she served as senior vice president of research and development (R&D) for Nissan North America Inc. Carla was responsible for vehicle engineering and development operations in Michigan, Arizona, Mexico, and Brazil, managing a $500 million budget and 2,500 employees. In this role, she improved the efficiency of Nissan's R&D functions.

Terry Barclay
President and CEO
Inforum

- Tel: +1.313.567.0232
- Email: terryabarclay@gmail.com
- www.linkedin.com/in/terrybarclay
- Twitter: @terrybarclay
- Topics: Women leaders, leadership development, gender-balanced businesses

Education

- BA, College of Wooster
- MSW, University of Michigan

Terry Barclay is a leader in building gender-balanced businesses with over 35 years experience working with top management teams of some of the world's best-known companies to capture the benefits of inclusive work environments. As president and CEO of Inforum, Terry oversees the only professional organization in Michigan—and one of only a few in the country—that combines strategic connections, proven professional development programs, a respected forum for new ideas, and original research to accelerate careers for women and boost talent initiatives for companies. Under Terry's leadership, Inforum has become a trusted ally and sought-after resource in helping companies advance gender diversity to find and develop future talent.

Inforum's research-based leadership development programs have been widely used by automotive companies for current and future women leaders. Those include "Ascending to the C-Suite," a collaboration with the Ross School of Business at the University of Michigan that helps senior-level women acquire the specialized knowledge whose mastery is critical for C-suite executives.

Inforum also publishes the biennial *Michigan Women's Leadership Report*, which tracks women's leadership in the state's publicly traded companies.

Inforum is active in the automotive industry through its AutomotiveNEXT initiative, a forum where automotive industry leaders can connect and collaborate on innovative ways to attract and retain top female talent. Inforum also offers inSTEM, an initiative aimed at closing the STEM gender gap by engaging more women working in STEM fields in local STEM programs.

Terry serves on the boards of Cranbrook Institute of Science, The Nature Conservancy of Michigan, and Rebel Nell LLC. She is an active angel investor through Michigan Angel Funds I and II.

acknowledgments

Writing a book can sometimes be a solitary enterprise, but publishing one isn't. And so this is where we get to thank those who helped make this book a reality.

First and foremost, we want to thank the women who took the time to provide thoughtful insights, useful advice…and amazing stories. The automotive industry can be a tough business. Rapid changes in technology, consumer preferences, and the economy demand strong and nimble leaders. The women in this book are some of the best of those leaders, and we could not be more grateful to them. Their commitment to nurturing other women's careers is humbling.

Our co-producers, Sherry Nigam and Dan Reilly, and many others behind the scenes at SAE provided much-valued patience and guidance as we learned the process of book creation. Thank you.

We also want to thank our coworkers, who helped us refine our ideas at the beginning of this project and helped with proofreading at the end. A special nod to Inforum Vice President Cindy Goodaker, who was instrumental in developing the questions used in this book and provided fact-checking and other editing assistance.

Lastly, we want to thank you, the reader. We truly want to make a difference for women in automotive and help create a way for all people—women and men—to be successful in our business. We hope you feel as inspired by the women in this book as we did.

Carla and Terry

The competition for talent has never been tougher. As industries are increasingly technology-based, the challenge for the automotive industry is to be both visible and accessible to a diverse range of potential team members and leaders.

And even though two out of three car-buying decisions are made by women, the number of women working on automobiles is quite low. This has never made sense to us—why would we have a demographic designing, testing, manufacturing, and selling a product that doesn't represent the majority of buyers? One reason is that the path isn't always apparent to talented women. To this end, this book presents the career journeys—in their own words—of women who have been, and continue to be, successful in a variety of automotive careers.

Each of the 34 women in this book was asked to select one of several questions in six categories of experience: education and lifelong learning, work-life integration, mentor and sponsor relationships, taking charge of your career path, resilience, and personal satisfaction. Biographies of each woman are included.

As we read their responses, we were in awe. It became crystal clear that the "road to the top" is a unique experience, and there are many paths to this success. There were many different ways to manage difficult situations, career changes, work-family integration (if this even exists), and others. There is no set pattern—each woman forged her own path. You will see that every individual's experiences were different and their expressions of these experiences are equally different. Responses are published as written to express their voices.

There were some common threads—the importance of sponsors and mentors, taking risks, and resilience—but also a great diversity of experience.

As automotive segues to mobility and needed skills change to customer experience, human behavior, and artificial intelligence, it is more crucial than ever to have a workforce where all voices can be understood and heard. Diversity of thought is critical.

Careers in the automotive industry can be more challenging than many, but strong and empathetic leaders make a difference—you'll read examples of this throughout the book. As one woman leader wrote, resilient leaders "give themselves grace, and they give others grace."

We hope you enjoy reading this book and that it can help you make a difference in your career. Become a mentor or sponsor for the young women in your companies. Further, use these answers to be a mentor/sponsor for the talented women whose careers you are working to advance. Become a role model for girls to give them confidence that they can and will succeed in automotive or any STEM career they choose.

Spread the grace.
Carla and Terry

Sue Bai

Principal Engineer
Automobile Technology Research Division
Honda R&D Americas Inc.

Sue Bai is a principal engineer in the Automobile Technology Research Division of Honda R&D Americas Inc. Bai's areas of research include wireless communication for in-vehicle navigation systems, telematics system design and development, and connected and automated vehicle system research. She currently leads a team that supports Honda's transportation safety and mobility goals through connected vehicle and V2X communication systems.

Bai has held leading roles on SAE V2X technical standards committees for many years, working to improve safety and mobility for a variety of road users including vehicles, pedestrians, cyclists, and road workers. She is also the Honda technical leader for various industry-government collaborative projects including Ohio's 33 Smart Mobility Corridor that, when fully operational, will be the longest stretch of continuously connected highway in the world.

Education and Lifelong Learning

How important is it for companies to create lifelong learning opportunities for their employees?

As technology progresses and pushes the automotive industry, learning new skills becomes critical in order to adapt to the changing landscape. It is great to know that Honda cares about my growth through providing education assistance. In return, I get to apply what I learned at school to improve my work.

I also think my job provides me with opportunities for informal learning as well. I enjoy talking with Uber or Lyft drivers. Often, the 20-30 minute ride provides me with a chance to learn how the drivers like their jobs and what challenges they have as part of the ride-sharing industry. I also like to hear how the new transportation system affects their life.

These ad hoc interactions give me tremendous knowledge on society's perspective regarding what we are doing in the industry. It is often eye opening to me.

Work-Life Integration

How have you dealt with work-life integration in your own career?

This may be unusual, but I really enjoy working. My passion is to deploy connected vehicle and infrastructure technology, break the isolation of each road users, and empower people to help make transportation systems safer and more enjoyable. My work is my life and they are fully integrated.

Mentor and Sponsor Relationship

Do you benefit most from mentor/sponsor relationships or from other relationships or networks?

In Confucian philosophy, there is a saying that states "Among three people walking along with me, at least one is my teacher."

I consider most, if not all, of my coworkers as my mentors in a sense. Some are coworkers whom I learn from every day. Some are people I meet at events or during nonwork-related occasions who teach me about nonwork topics. I once met a gentleman at the beach when I was taking my morning jog. He was picking up trash to clean the beach. We started talking and he taught me much about the recycling industry in which he worked, and eventually, we became friends.

The more serious mentoring for me often comes from senior executives (including executives who have retired) that I work for or had worked with. I believe it is important to have a structured system and synergistically generated mentorship. So far, my mentors belong to the latter category and I am very fortunate to have the opportunity to interact with higher-level people from various companies.

However, I appreciate that Honda also has structured programs that allow associates to connect and learn from leaders.

Taking Charge of Your Career Path

What intentional decisions have you made about your career and were there opportunities you received that you had never considered?

I have made conscious decisions about my career based on advice from our organization's former president who happens to be one of my mentors: Career paths are results, not goals.

His advice made a significant impact on my career decision. We discussed that the job needs to be interesting and exciting; innovation should be valued, encouraged, and supported throughout the organization. In addition, the company's value and vision shall have societal benefit, not just be used solely for profit, and the employees must be treated with respect that they earned through hard work. For me it has been about finding the right opportunities and having the right environment to empower me to do my best for what I believe is right.

Certainly, we all wish to be promoted; however, that is a reward, not a requirement.

Resilience

What qualities make up resilience in a leader?

I think a leader should start by being a good listener. The leader's job is not to ask people to follow, but instead to make people want to follow. You should support the team with necessary resources and clear direction, and protect the team when needed. To be resilient, a leader should believe in the value of the project/products they are working on, be open minded, listen to other people's opinions, learn from others, be honest when mistakes are made and then learn from all of that, and make one's team even stronger team.

Personal Satisfaction

What would you say gives you the most satisfaction in your career?

To be able to work on life-changing technologies and improve society. I'm proud to think back at how I have grown from an entry-level engineer to an engineering lead working on large-scale research projects that expand to external partners such as universities and government agencies. I am also pleased to have learned from people I work with, and as a result, I think they have made me a better person. It is also satisfying to be valued and recognized by the organization for my work and my team's effort.

2

Susan Brennan

Vice President
Chief Operations Officer
Bloom Energy

Susan Seilheimer Brennan has 25 years of experience in global manufacturing and operations for the automotive and energy industries, with strategic leadership roles in Fortune 100 companies Nissan Motor and Ford Motor. She is one of the few women who have held an executive position at a leading Japanese corporation. Her unique experience includes spearheading several large-scale initiatives within complex, global organizations aimed at addressing financial challenges. Brennan drove the companies' transformations with systemic process and corporate culture change, including managing international teams through difficult business transitions that led to improved operational and labor efficiency.

Brennan is an experienced independent nonexecutive director to the board of Senior plc, an international, market-leading, engineering solutions provider with 33 operations in 14 countries headquartered in the UK. As chief operations officer for Bloom Energy, Brennan is the chief executive responsible for all new product launch strategy and sales execution, operations, capacity management, global supply chain, purchasing, labor management, EHS strategy and compliance, and government affairs. This includes providing leadership in revenue, sales, and bid delivery and contract negotiations, with shared accountability with the chief technology officer.

Brennan received her MBA in economics from the University of Nebraska at Omaha and her BS in microbiology from the University of Illinois at Urbana-Champaign, and she is an alumna of the Prince of Wales's Business and Sustainability Program.

Education and Lifelong Learning

How important is an MBA or other graduate degree?

I find it interesting that, over my career, I have been asked this question multiple times and I always say "yes." I am from a generation where the question would not be asked, it was assumed if you wanted to distinguish yourself and move forward in leadership, an MBA or master's in your field was necessary. For those of my generation, it separated you from others who were ambitious and talented. It could be used as a tie breaker if all other qualifications were equal, but mostly it demonstrated that you were committed and focused and could take on additional challenge and be successful. It was a marker of drive, ambition, focus, and commitment. It gave a company confidence you could and would start something, finish it, and go above and beyond. Now that I am in Silicon Valley, working with Gen X, millennials, and Gen Y, I still give the same answer.

Here, I am in a world when a PhD is the standard. No one goes by "Dr," it is just assumed. The advantage of an advanced degree in today's world is that it helps you understand process and provides a broader perspective. With the rapid pace of technology and increased complexity, you need the fundamentals of chemistry, math, and physics, and you need the tools to apply them to an ever-changing landscape of technical innovation and consumer demand. An advanced degree that focuses on "how" to use the fundamentals and, more critically, one that helps you build a network of like-minded people is crucial in today's world.

I see often events in SV where Wharton, Harvard, and other elite MBAs come together to network and use each other's knowledge to stay relevant and even work together to advance new ideas and innovation. In my day, universities gave you a diploma and then hoped you might donate a building someday. Today, they have strong networks that keep graduates linked and provide lifelong learning through alumni activities such as seminars and networking connections. The days of the "happy hour" are long gone. They are now serious events where major topics are discussed with alumni and guests, if you are fortunate enough to get invited.

In summary, the answer is a resounding *yes*. The reasons may not be the same as when I received my MBA—many of the strategies are irrelevant or less relevant today, but the "tactics" and fundamentals continue to serve me well. Today's advanced degrees that are relevant will be more tool, leadership, and networking focused and will support the graduate in a lifetime learning environment and that alone is worth getting an MBA.

Work-Life Integration

Manufacturing continues to cause a divide in gender—why is this? What can be done to support more diversity?

This is always the question where I get "booed." I have not and don't have work-life balance. Early on, it was a choice of circumstance. I grew up without means and I was driven and focused, and made sacrifices I probably did not need to make, but the combination of being the only woman in the room and my sense of worry over money, even when I was financially stable, has caused me sacrifice work-life balance. I partially blame my roles—operations, supply chain management—and I partially blame myself. I went back to work immediately after having a baby, and moved with a 6-week-old, because I had received a promotion and did not want to lose it. If you put it into context, I was the first ever pregnant plant manager at Ford, you may excuse my paranoia, you may judge me, or you may salute me. The choice is up to you.

Ok, so what to do with the 800-pound gorilla in the room? Do I give today's woman advice to ignore the unintentional or intentional bias that comes with being a woman in manufacturing? The things that can't be changed: shift work, launches, shutdowns, changeovers, and the almighty production report.

So, what to do with it. A woman who chooses to go into manufacturing must have tools to support her throughout these times. Find a "buddy"—you cover their zone for a soccer game while they cover you for a ballet recital. Build your team—prepare and assign a team leader, with your boss's permission, give them the opportunity to cover for you when you need a Friday off or need to go into a daytime parent-teacher conference and need the sleep for one shift. Put them in a position of being able to apply for the next job bid and assure that one of your peers will check on them while you are out.

Many women have been successful in manufacturing, but it takes skills, tools, partnership, finesse, and stamina. If you want to be "one of the boys" that is up to you. But if you want to acknowledge the fact that you carry more of the burden for the household, childcare, and that, last inviolable, pregnancy, find a company that is willing to work with you to find solutions. It can be done. However, you have to advocate strongly for yourself because you are already several steps behind being a woman in this field (one person's opinion).

Mentor and Sponsor Relationships

Do people understand the difference between mentorship and sponsorship? Is it a part of regular succession planning in your companies?

I fundamentally believe that this definitional difference is one of the top reasons why women continue to be challenged in promotions and workforce progression. I believe that well-intended mentoring programs end up making women look "weak" and in need of help when what they need is someone they can trust, with experience to gain an opportunity to get perspective and "think out loud." Mentoring is important but often misunderstood and misapplied. Mentoring is something that should be organic and based on trust and common ground—it needs to be someone you are comfortable with, able to speak to, should not be in your chain of command and, if possible, not in your company.

I want to be very clear—every woman who wishes to progress in any industry, but in an industry as complex as automotive, must have a sponsor. You must have someone "in the room" advocating for you. There must be someone, in a position of influence, that knows what you are working on, will give you unfiltered, unvarnished advice, and will put your name in the hat and fight for you when the time comes.

If anyone reading this book does not take away any other factoid or learning, learn this—*go get a sponsor*. Make sure they are really a sponsor. I have sponsored several women in my career (some with better success than others). Why is the sponsor critical to your success? When the question gets asked, "the job requires the person to move and I don't think Sally will move." The sponsor responds, "First, I know for a fact she is willing to move for a promotion, and we should be asking her directly, instead of making the decision for her if she is the most qualified candidate." Very different outcome if the person does not have sponsor in the room. Or, this job would be great for someone who is high potential but does not have an engineering or manufacturing rotation. *Ask*: Who on our list of high-potential employees meets this criteria? If you have a sponsor in the room, I argue, and my experience supports it, you are more likely to be on that high-potential list and be considered successfully.

How the conversation *can* go without it—"you're right, Sally just had a baby and won't likely want to move so let's give it to Bob. We can consider her later, *if* she decides she wants to be mobile." Or "I agree Sally meets the criteria for a cross-organizational rotation but engineering and manufacturing have many demands and I am not sure she is ready to take them on"—with no disagreement, the job goes to someone else. How do you change the conversation—make sure you have someone in the room that knows you, trusts you, is pushing for you, and will advocate for you. Over the years, many of my sponsors have been men. I often argue this actually helps a woman to have a male sponsor. In a world of bias, unconscious or not, my experience is a male sponsor can carry additional weight, especially in automotive.

Taking Charge of Your Career Path

If you changed companies, what was the compelling reason and was the move beneficial? Why?

Second nugget that I hope readers take away from me:"Disrupt yourself before someone does it for you." What does that mean? Either you can be in the "driver's seat" or you can be a passenger with no control on where you or your career goes. If I fundamentally look at what drives me, it is loss. My dad died at 29. Each day is a gift. If I don't go out there and try something, I will regret not trying and my number 1 fear is "fear of regret." The worst I can do is go back to doing what I did before. The best that can happen is that I meet state leaders, travel the world, get invited to RC Cola and Moon Pie parties, meet some of the most incredible and genuine people ever, change the world, and, if I am really lucky, build a better future for my children and my nieces and nephews. To paraphrase Gandhi and Margaret Mead and two of my favorite mantras: "Be the change you want to see" and "never underestimate that a small group of people can change the world; in fact, it is the only thing that ever has."

You don't need to have your name in lights (it is fun though), but you need to know that you are making a difference and doing something that matters. That can be providing a job for someone who otherwise might not have access to a good job or you can build electric cars and bring mass transportation change to the world. You can receive rewards, or your reward can be the satisfaction of a job well done.

I don't think that everyone has to follow my path—you may go from trim to chassis or engineering to production, disrupt yourself, put yourself out of your comfort zone before you become complacent and change happens to you, instead of your driving the change.

I changed because I had an opportunity. If one is open to opportunity, you will find, the opportunities find you… Change is hard, change is scary, change with kids is tough because you may be disrupting them at very vulnerable times in their lives. This is the only time I would suggest being moderate in your change. One needs to keep growing but the growth may require taking on a role at the current level, in your current location, in order to accommodate your current circumstances.

Resilience

Do you consider yourself resilient? If so, how did you become that way?

Resilience is earned and learned. It is when your dad dies when you are 5 and you cannot shake that day no matter how many years pass, it is when you find out that the college fund you were promised doesn't exist and you are six months away from college without a plan, it happens when your boss at a major automotive manufacturer humiliates you by making you carry a stuffed monkey while on the job (that will be one subject in my book) and you soldier on with poise, grace, dignity, and little bit of ass-kicking spirit. All of the above have happened to me, and so much more, but I am still standing.

Resilience is truly, I believe and I have had confirmed by many people I respect, one of my top qualities. There are many days I wish I had the ability to walk away and never look back. But like all of us, I have bills to pay and children to raise and goals of my own to meet. Plus, it is always rewarding when you win an unfair fight, and I have been fighting those my whole life.

Don't expect to win all battles—resilience is knowing which ones to pick. Resilience is also the ability to lose with dignity while keeping your dignity. I believe in Karma so those people who impact you, they will get their payback, it will take too long and they will likely be successful longer than they should be, but everything evens out in the end. When situations challenge you: learn, learn, learn… ask yourself "what can I learn to make me a better person, a better leader."

When I first came into the workforce, men would come to me and tell me, to my face, they did not want to work with a woman. When I tell these stories today, today's young women are shocked. I was grateful. I had full knowledge of the environment I was in and where people (in this case the men I worked with and around) stood with me. To me, this knowledge was power. I had a boss try to undermine me by falsifying quality data on vehicles I was responsible for to try to get me fired. Not only was it expensive for the company to chase after repairs that did not exist, it could have been a safety issue for consumers. However, because I held myself with dignity and did not try to be someone I was not, one of the union workers came to me and told me what was happening. I was then able to address the issue directly. This union worker was someone who likely did not want to work for a woman but, in the end, wanted to do what was right because of his own values, regardless of my gender.

Resilience, in my definition, is the ability to have the dignity to withstand the "witch hunt" while retaining my composure and continue to show up to work every day knowing that I was in a very hostile environment. I did not let it change me. These experiences may have been painful and exhausting but someone has to pave the way. It might as well be me.

My most important tenet—you cannot control what those above you do but you can control your own actions. Never behave in a way that diminishes you regardless of what others have done to you.

You may find yourself in extremely challenging and tragic situations: you may have a significant financial event, your company may close, or you may find your child has a terminal illness. Even through these unthinkable situations, stay focused and positive. If you have built a good network, lean on them to help you through the challenging times.

Never despair—be creative, create a mantra, and push forward.

A little secret—anyone who has ever worked to discredit or impact me in a negative way has always made me stronger and led me to a better path… wish it had not been so painful but it is better than having someone of lower moral character and values kill the special spirit that makes you the leader you are. Without resilience, in my opinion, the best leaders get taken out of the workforce and that diminishes all of us.

Personal Satisfaction

What would you say gives you the most satisfaction in your career?

I get the most personal satisfaction in my career when I see something in someone that they don't see in themselves and persuade them to act on it: when a production worker applies and becomes a supervisor; when a supervisor becomes a manager; when someone who was told they were not smart enough goes to college and gets not just a BA but, even better, a master's; when people see the amazing individuals that they are; when they bring their "whole selves" to work.

Money and position are nice and having a view of 49ers' stadium and the San Francisco Bay are certainly nice perks, but they don't compare to the satisfaction of knowing you changed a life and started a circle of success by enabling people to see who they are. These transformations impact families, communities, and generations—I find it to be one of the top elements of satisfaction for me, maybe going back to my Puritan, Midwestern roots.

In full disclosure, one perk I do miss though, is a free car. I received much satisfaction from my time in automotive from driving amazing vehicles. Especially, before they are launched and people follow you and flag you down to see what you are driving. If you wonder if you should go into automotive for personal satisfaction, drive a Ford GT or Nissan GT-R just once. It won't disappoint.

3

Kim Brycz

Senior Vice President of Global Human Resources
General Motors

Kimberly (Kim) J. Brycz was appointed to the position of senior vice president, global human resources (HR) on March 12, 2018. In this role, she leads an HR team and systems that build enterprise-wide employee engagement, develop talent, and support strategic planning at all levels.

Previously, Brycz served as executive director, global product purchasing where she oversaw General Motors (GM)'s $80 billion annual global product spend as well as customer care and aftersales purchasing. Brycz was instrumental in transforming supplier relationships by providing strategic solutions to future sourcing and supplier engagement.

Brycz, a native of Detroit, began her GM career in 1983 with the Cadillac Motor Car Division in Detroit. Since then she has held various positions in GM's global purchasing organization, including the global purchasing lead for electrical systems, batteries, and hybrids and interiors. Prior to her most recent role, Brycz served as executive director, global purchasing, indirect materials, machinery, and equipment.

Brycz received her bachelor of business administration degree from Michigan State University. In 2010 and 2015 she was named one of the 100 Leading Women in the North American Auto Industry by *Automotive News*. She is an active leader on two of GM's employee resource groups, participating on the executive boards for the GM Asian Connections and GM Women Group. Additionally, Brycz has served as the GM executive champion sponsor for Making Strides Against Breast Cancer.

Education and Lifelong Learning

How can someone best improve their strategic thinking skills?

To improve strategic thinking, it is imperative that employees connect themselves with their company's vision and mission. Employees better understand the importance of their work in relation to the business enterprise when they are able to connect their responsibilities to a sense of purpose.

To gain a better understanding of one's purpose in an organization, I always advise my mentees to raise their hands for special projects. Be vocal and proactively seek out projects that allow you to experience how the larger company operates. Observing an enterprise holistically and gaining a better understanding of its operations allows you to achieve a greater strategic perspective.

We must also be cautious about focusing on achieving individual metrics. They often drive short-term, one-dimensional results that tend to be less strategic. Because of this, leaders must encourage decisions that impact long-term, corporate-wide results, not just quick organizational returns. To reinforce the importance of strategic thinking and decision-making within a company, leaders must recognize and reward behaviors that are innovative and collaborative.

Work-Life Integration

Have you ever chosen work over family or vice versa? Has this gone well? What did you learn?

I must confess that I have chosen work over family on occasion. When I reflect on those moments, however, I question if the company was better off with my decision.

There is a balance of personal duty, guilt, and responsibility that is associated with making these decisions. Most of the time, my sense of guilt was not driven by my boss but more so by the need I felt to prove that I can manage all aspects of my life seamlessly. To the contrary, as hard as I tried to balance being a mother, wife, and professional, there were times when some aspects of my life suffered in sacrifice of others.

On one occasion, I recall my boss at the time approached me in an effort to detect the reason I seemed to be struggling with balancing it all. I finally admitted to him a critical pain point of mine during that time. I had three sons, all under the age of 5 years old, who required separate drop-offs at different day cares, and it was costing me almost my entire salary! I enjoyed my job and valued my work, but the effort did not match the reward I deserved to provide for my family.

So, he immediately worked on a remedy and offered me a pretty significant raise that would help cover day-care payments. He also offered that I work from home once a week to be with my boys—during a time when technology didn't make working from home as effortless as it is today. My leader understood that these solutions, which better balanced my personal life, would encourage me to give 150 percent at work.

The reality is that life throws you curveballs, but family is the foundation. Great leaders realize this and will support you because they know your productivity is impacted by your sense of balance. Often, finding a solution to your pain point is as simple as speaking up.

Mentor and Sponsor Relationships

Did you benefit most from mentor/sponsor relationships or from other relationships or networks?

I benefited from both mentorship and sponsorship in different ways throughout my career. A true, effective sponsor is someone who knows you both personally and professionally and has first-line experience with your capabilities and leadership qualities. A sponsor will keep you in the forefront when key decisions are being made. He or she may also double as a mentor from time to time to advise you on growth opportunities and improving your blind spots. I have been very lucky to have people in my life who truly care about me provide helpful advice and encourage developmental opportunities.

The greatest lessons sponsors have taught me were the powers in having confidence in myself and realizing the value of my skill set. As an introvert, I would not typically promote myself or conscientiously highlight my key strengths and accomplishments. Luckily, I have had many sponsors throughout the years—both men and women—who have played that role on my behalf and encouraged me to advocate for myself as well. I am humbly grateful for that.

I now pass these lessons on to the individuals I mentor and sponsor. The men and women who have helped me further my career have also afforded me the ability to now pay it forward to future generations of leaders. I can help instill the confidence I did not possess growing into my career and offer an encouraging voice to let them know, "You are ready, and you can do this."

Taking Charge of Your Career Path

What would you do differently about planning your own career if you were starting over?

I approached each step of my career conservatively, cautious of the opportunities I would commit to. I worked hard and was always very confident about my work at hand. The problem was I was not always as confident in taking on that next role. I did not proactively raise my hand and say that I was ready for that next challenge. This was mainly because I always wanted to improve upon what I was doing at the moment.

The most dramatic change in my career arose from my recent leap from executive director of global product purchasing—a part of GM's business that I have supported my entire career—to becoming senior vice president of global human resources. When first offered the position, in typical Kim fashion, I said that I would need the weekend to think about it as opposed to immediately pouncing on the opportunity. I was nervous to lead an entire function I had no previous experience with.

However, in speaking with an influential leader within the company, it became apparent that so many of my newfound job responsibilities would mimic the great skills I have worked so hard to strengthen throughout my career thus far. My skills were transferable; only the content of my work would change. In that transition, yes, there was a learning curve and, yes, there were things I would have done differently looking back. But I have never once regretted the jump to senior vice president.

The bottom line is that leaders challenged me more than I challenged myself. I took on roles that I thought I was taking on too soon. I even declined opportunities because I felt I was not prepared. So, if I had a do-over, I would place more confidence in myself and my ability to take on stretch assignments.

Resilience

How do you build resilience in a team?

Given the ever-changing dynamics of the auto industry, it is important to keep teams focused on the wins—both present and future. We need to encourage them to seek opportunities, possibilities, and learnings from situations that may, at first, feel problematic.

During times of crisis I have found employees in a sort of frozen state of chaotic panic. My remedy always begins with instilling organization.

The first phase requires the team to step back and assess the situation. Second, we must help employees understand their purpose within the larger organization and confirm the importance of their contributions. When employees see their work is adding value, they can better manage ambiguity and change, and they tend to focus their efforts more efficiently. Helping people look at situations from varying points of view allows for broader understanding of the problem and an increased likelihood of forming creative, successful resolutions.

Ultimately, I always counsel people facing challenges to break down the issues, identify purpose, develop structure within the approach, focus on key initiatives, and always celebrate the wins.

Personal Satisfaction

What would you say gives you the most satisfaction in your career?

In my career, I've found the most satisfaction when vision, strategy, and a team's ability to execute plans are in perfect harmony.

As a coach, I feel my role is to not only deliver results but to guide employees to understand the mission and future impact of their work. This connection between effort and execution propels emotional attachment to the results. And there is no greater feeling than to share in that win with peers who have given their best. When this occurs, it is inspiring to see this all come together and reach the collective goal.

4

Shari Burgess

Vice President and Treasurer
Lear Corp.

S hari Burgess is vice president and treasurer of Lear Corp., a role she has held since 2002. Ms. Shari is responsible for understanding and providing the long-term liquidity requirements to navigate economic cycles, support a multicountry growth strategy, and deliver a competitive return to shareholders. Shari leads a global team in charge of capital market strategy and execution, daily cash and related cybersecurity systems, foreign exchange and interest rate risk management operations across 39 different countries and regulatory regimes.

In addition to being treasurer, Shari also served as Lear's chief diversity officer from 2014 to 2018. During this time Lear increased the minority representation on the board from 10% to 30%, increased female leadership globally by 200%, attained a Human Rights Index rating of 100%, and achieved membership on the Billion Dollar Roundtable.

Ms. Shari also serves as the chair of the Health Alliance Plan (HAP) board of directors and is on the board of Henry Ford Health System (HFHS), together a $6 billion integrated health care system. Shari is also an Inforum board member and had previously served on the board of the Michigan Roundtable for Diversity & Inclusion.

Ms. Shari obtained her CPA while working in audit with Ernst & Young LLC and was also the corporate controller of Victor International before joining Lear. She received a bachelor of arts degree in professional management and economics from Albion College and a master of business administration degree from the University of Michigan.

Education and Lifelong Learning

How have you structured your own approach to lifelong learning?

Throughout a majority of my career I was not in an organization that provided leadership training or job rotations. I was also in finance roles that are often incorrectly perceived as specialized. So, I had to create my own learning opportunities.

Being a self-starter is important in business, including with respect to your own professional development. I always challenged myself to seek projects that took me out of my comfort zone. I volunteered for extra assignments considered nontraditional for someone in my role, assignments that drew from my knowledge but crossed over functional lines. Every chance to work with professionals within and outside of Lear that I did not have a reason to interact with on a normal basis broadened my business perspective and deepened my understanding of the industry and my company's role within it.

Working in a cyclical industry like the auto industry is never boring. Rather than exiting the industry in difficult times, my professional growth was accelerated from stepping up to the challenges. Understanding what went wrong and why is by far the most valuable learning experience; it strengthens one's professional judgment immeasurably. In addition, the character one demonstrates during times of adversity is the most important and longest-lasting impression one can create across the organization and throughout the community.

I have also sought external learning opportunities by not just joining but actively participating on various charitable boards and a nonautomotive corporate board. Perspectives gained from other organizations, industries, and board members have enabled me to approach my daily Treasury role much more strategically.

Work-Life Integration

How have you dealt with work-life integration in your own career?

My generation was taught that women could have it all, and we can … just not always at the same time. The trick is to embrace your career and family choices and not feel guilty. Always be present, be fully engaged when you are at work and enjoy being with your family when you are at home.

There were natural points where I needed to make a push in my career due to time-sensitive transactions. Besides learning to delegate, I had to learn to prioritize and let the little things go, at work and at home. My house was hardly ever perfect, I was not on a first name basis with all my kids' teachers and my adult sons are just now realizing that I *can* cook. But through the chaos, my children always knew they came first. I was there for the important things, both big and small. I helped with homework (sometimes by phone) and supported them in sports (even if I was the only mom at the hockey rink in a skirt and heels). This often required me to go back to work after the kids were asleep. But truth be told, I often achieved more critical thinking in the uninterrupted silence of night than I did all day at work.

There were also times I had to put my family first. I passed on overseas assignments and turned down external job offers that could have accelerated my career but required a move at an inconvenient time for my children. It may have taken me a little longer to move up the ladder but the family balance was worth it.

Mentor and Sponsor Relationships

How important have mentors and/or sponsors been in your own career? Have they been men or women?

Although I have never had a formal mentor or sponsor relationship, there were many men and women throughout my life that have provided an invisible hand of encouragement or opened doors that shaped my career.

Throughout my education several teachers took an interest in my potential. They put in the effort to develop different assignments and set higher expectations for me than the rest of the class. Although I did not appreciate the extra work at the time, I now understand they were instilling in me the confidence to challenge myself intellectually at a time when it was not popular for women to be smart.

Many business colleagues have generously shared their time, expertise, and experiences while modeling leadership qualities daily. Informally, they taught me that a leader takes ownership when things go wrong, shares the credit internally and externally when things go well, provides flexibility when needed, and does not avoid uncomfortable discussions when supportive, constructive criticism could help. Everyone is unconsciously either a positive or negative mentor to those paying attention.

My sponsors, on the other hand, include external business advisors and executives I have worked with on charitable and other boards. They have unselfishly raised my profile throughout the business community by inviting me to speak at functions, sharing their strategic connections, and acknowledging my contributions to my management team. Being endorsed by senior management's own peers throughout the community increased my professional credibility and allowed me to be viewed in a completely different light—more as a peer.

Taking Charge of Your Career Path

What intentional decisions have you made about your career and were there opportunities you received that you had never considered?

My path to becoming the treasurer of a Fortune 150 company was not traditional. To be honest, I never had a career game plan in mind. I was conscious, however, about making choices that provided me a depth and breadth of skill sets that would keep my options open.

My undergraduate major was in accounting, a practical degree that was the most likely to lead to a job. To hedge my bet, however, I double majored in economics. I worked as a CPA for several years but ultimately decided that I wanted to be on the front end of the decisions, negotiating the deals as opposed to accounting for them after the fact. The analytical skills I learned as a CPA turned out to be invaluable though. Knowing how a transaction would be accounted for provides a competitive edge in evaluating the true economics of a deal. And my audit and accounting experience ultimately qualified me to be on the Audit Committee for the board of a large hospital system.

After going back for my master's degree, I joined Lear, at the time a relatively small but dynamic privately held automotive supplier. On the path to becoming a Fortune 150 company, I got to travel the world, participate in multiple acquisitions, finance all the deals, take the company public, perform financial planning and analysis, and work directly with operational finance, shareholders, Wall Street, and the board. It was not one decision but the breadth of perspective and experience gained over time that qualified me to become treasurer—a role I never considered until I was at Lear.

Resilience

What qualities make up resilience in a leader?

Adversity and change are a fact of life, in business, and within our personal lives. I have heard it said that the only constant in life is change. The pace of technological change and geopolitical influences in the automotive industry makes this even truer today than ever.

A resilient person can recover quickly. A resilient leader looks for the lessons learned and the opportunities created by change or adversity and acts to adapt and move forward to a position of even greater strength. If anything, they become even more determined to succeed.

Adversity comes in many forms: economic/market disruptions, mistakes, failed projects, lost jobs, or internal politics. I consider resilience a choice. For some resilience comes easily, and others must actively think about it. Either way, resilience is the key to maintaining one's energy, motivation, and positive attitude.

My approach to resilience is to

- Remain confident—do not dwell on the negatives. Remember all your successes; don't lose your sense of self. If there was one right answer or strategy, we would all know it!

- Focus on the things that matter and the things you can change—do not waste time, energy, or emotion worrying about things you can't change or influence. Reframe the issue to allow creative problem-solving or opportunity identification.

- Remain optimistic but realistic—be willing to learn from failure or adversity and adapt or change course.

- Take action—fear and disappointment are natural but inaction is a choice; focus on how to constructively move forward.

Personal Satisfaction

What would you say gives you the most satisfaction in your career?

My best piece of advice for young women is to define your own success. There is no one timeline, job title, or salary that determines success. Find what motivates you and do it on your own time frame.

For me, career satisfaction does not come from a high-profile job title. You seldom see a treasurer on the cover of *Fortune* or the *Wall Street Journal*, or if you do it is not for anything good. It usually involves a financial scandal—think Enron. I find career satisfaction from being able to contribute in a strategic way and having a positive impact. That is why I like treasury—negotiating and setting the liquidity and capital structure of the company has a direct impact on the ability of the company to meet its strategic objectives.

Becoming the treasurer *and* chief diversity officer was probably the most challenging role of my professional career, but it was also the most rewarding. Our CEO consistently says that our people are our most important competitive advantage.

Strengthening the talent pipeline and creating diversity of thought through accelerating the advancement of women and minorities globally not only enhanced Lear's economic value to shareholders, it has had a direct and positive impact on people's lives.

A colleague once told me that a job is only as big or as small as you make it. Raise your hand, get involved. Not all leaders have recognizable titles but they have significant influence on the organization, both directly and indirectly. I have declined positions with larger titles but less impact and have never regretted it.

5

Melissa Cefkin

Principal Researcher-Human Centered Systems
Alliance Innovation Lab Silicon Valley

Dr. Melissa Cefkin leads a social- and human-centered research and design team at Alliance Innovation Lab Silicon Valley, focused on the development of autonomous vehicles (AV) from the perspective of human-AV interactions and relations, and more broadly on the future of mobility.

She joined Nissan in 2015 from IBM Research where her research focused on work and organizational practices in the context of automation. She also has experience in design and consulting and was previously a director of Advance Research and User Experience at Sapient Corporation, and a senior research scientist at the Institute for Research on Learning.

Melissa was a member of the National Academies of Science, Engineering, and Medicine committee on IT, Automation and the U.S. Workforce (see their report here https://www.nap.edu/catalog/24649/information-technology-and-the-us-workforce-where-are-we-and) and served on the board of directors and as conference co-organizer for the Ethnographic Praxis in Industry Conference (EPIC).

She is the editor of *Ethnography and the Corporate Encounter* (Berghahn Books 2009) and numerous other publications. A Fulbright award grantee, she enjoys frequent presentations at conferences, has served on numerous editorial boards and program committees, and is frequently sought out by the press for comment on issues that surround people and automation.

Education and Lifelong Learning

How important is an MBA or other graduate degree?

I have a PhD in cultural anthropology. This is not a typical advanced degree seen in the automotive sector! And what's more, I am here *because* I'm an anthropologist and I use this degree regularly on the job. I am certain I would not be working in this capacity without an advanced degree, nor without prior work experience at the intersection between human sciences and technology development.

Whether and what kind of an advanced degree is necessary or useful in automotive comes down to the specific job and function one works in, and often the details of the work itself. (Having previously studied people in their workplace, I've observed this is not unique to automotive but is true of any enterprise work.) An advanced degree is not always needed. I have built up a broad experience by working in research and innovation divisions and functions. In those functions it is common to bring specialized knowledge and skills to bear on the work. So having an advanced degree in a relevant field is extremely valuable.

More generally, knowing what counts as a valid question, how to formulate good ways to move forward, and knowing how to assess inputs of all kinds are things common to many jobs. Advanced degrees can help train and deepen a person's abilities on all those fronts. Moreover, while advanced education often focuses on specialization, at the same time pursuing an MA or PhD thesis also requires examining the world from various angles. This can be useful in any job and any function. That said, regardless of educational background, all workers today have to continually adjust to new knowledge and new ways of working.

Work-Life Integration

How have you dealt with work-life integration in your own career?

In my view, neither the notions of work-life balance nor work-life integration quite speak to the situation we find ourselves in. I think what is really meant is *job*-life balance and *job*-life integration. Whether self-employed or a member of a firm, people's work activities are structured by their position in support of the organization's goals. It is the position and the organizational dimensions that most require balance or integration. Work itself, in contrast, is a way to explore, to gain a sense of self-worth, and to engage with the world. At its best, it is a prime space for learning about the world and ourselves, and achieving a sense of accomplishment. Very often when it is less than that, the trouble is often more with the position and organization and less with the work itself.

I take pride in my strong work ethic and ability to plug away even in the face of adversity. And I strive to live a life of friendship, joy, and fun outside of my job. So when it comes to these notions, call me old fashioned, but I'm more on the side of work-life balance. I remain cautious about work-life integration. To be clear, that people's personal lives blend into their time on the job, and vice versa—for instance, personal time spent on phone calls or email—is perfectly natural. But, especially in these days of globalized work systems and collaboration, there is a risk that the job will continue to expand, squeezing out more and more of our energy and attention.

This is something I struggle with personally. In an effort to create balance, I try to follow some simple rules ("try" being the operative word.)

- No work email on my mobile phone.

- Avoid setting the expectation that email will be checked on weekends. I do often check my email, but I try not to send email. I don't want others to feel obligated to check on weekends.

- Mornings are mine. I start my day with personal reading—from emails to newspapers (the old fashioned kind when possible!) to books. Work stuff never comes first.

- I keep personal phone calls/texts to a minimum throughout the workday.

Mentor and Sponsor Relationships

Did you benefit most from mentor/sponsor relationships or from other relationships or networks?

I have found that my strongest support has come from people who know me through my work. As a cultural anthropologist bound for academia who landed instead in the thick of business and technology development, I have followed a rather nontraditional career path. My interests span the technical *and* social, academia *and* industry, and my commitments lie first and foremost with human-centered, pro-social concerns. And while I have not thus far been interested in following more traditional management tracks, I am interested in taking on new challenges and leading and guiding others. The people who have helped me the most—both as role models and for mentoring and advice—have been familiar enough with taking the odd path I've followed to get it, and they guide, advise, support, and challenge accordingly. Additionally, there's also been a lot of personal trial and error.

I believe profound insight and nudges can come from people one or two degrees removed from the immediate position, the quiet observers with whom contact may be infrequent. I've seen this to be the case, for example, in decisions about promotions and selecting people for new roles. It is not uncommon for it to be someone more distant to the person in question who offers a key insight. This reminds us that many people out there have useful contributions to make and insights to offer, whether identified through formal mentoring programs or not. And it is another reason to be deeply suspicious of the myth of the self-made man or self-made woman. Behind every successful person are people who've played roles in getting them to where they are.

Taking Control of Your Career Path

If you changed companies, what was the compelling reason and was the move beneficial? Why?

I have been lucky to gain from a few job changes during my career. Two are worthy of mention here. One was when I moved from a consulting firm, Sapient Corp., to an IT services and computer manufacturing firm, IBM. Both the role I left and the one I moved into were research positions. But I moved from doing research as part of our consulting efforts for clients, to an internal research division, developing service capabilities for

IBM to understand, use and possibly sell. I went from being customer facing to internal facing. I discovered that both kinds of roles have a lot to offer! Different challenges, different opportunities. The IBM position was also more focused on producing fundamental knowledge and IP and being a part of scholarly and scientific communities. But the main draw was that IBM at the time was a pioneer in a new area of science, service science. I found it very compelling to be a part of a new area of study, and one that was by definition multi-disciplinary.

The other move was when I switched from IBM Research to Nissan. Again, both positions were research positions, though in many ways the roles and work differed significantly—different time horizons, different funding and support, different industry contexts. Again, I was inspired to make this move to enter at the ground level of an emerging domain, autonomous vehicle development. More particularly, it was clear to me that cars, transportation, mobility, all deeply affect so many aspects of our lives in ways both big and small. Relationships of humans and machines are always interesting and important to examine, and ever more so in the context of a renewed focus on artificial intelligence. I am deeply committed to seeing that the human sciences be brought to bear at the very foundations of new directions in the industry.

Resilience

What qualities make up resilience in a leader?

One of my favorite quotes comes from the 1937 Frank Capra film *Lost Horizons*: "moderation in everything, including virtue." Even in the mythical utopia of Shangri-La, balance matters. Just like the High Lama of Shangri-La, resilient leaders engage deeply but also have the ability to keep moving and looking ahead. They avoid taking conflicts too personally, while at the same time view every challenge as a learning opportunity.

The automotive industry is in the throes of significant transformation. Its products increasingly vie for attention not just with traditional competitors but whole new industries. Not long ago imagining that an original equipment manufacturer (OEM) would not just be a customer of, but also a competitor to, the software industry would have been laughable. Now, with increasingly connected and automated vehicles, and the growth of the rich space of mobility services, it is reality.

My view of what makes a resilient leader is that they see things through a systemic lens. Recognizing that activities and events in other parts of a company, industry, or social arena can affect what is happening for their own organization and teams, helps accelerate preparedness and readiness to deal with change. Sometimes people may need time to absorb and work through what is changing, so allowing time for that is key, too. A good dose of humor never hurts either!

Ultimately, what I believe matters most to people, including myself, is to have a chance to do good work among good people. A resilient leader is one who recognizes and appreciates the joy of everyday achievements and experiences.

Personal Satisfaction

What would you say gives you the most satisfaction in your career?

Two things vie for giving me the most satisfaction in my work. One is that "ah-ha" moment when someone recognizes for the first time that something they took for granted is actually more complicated and worthy of a second look. The other is experiencing the smiles and laughter of my coworkers in their daily work. The second one is, I hope, self-explanatory. The bread and butter of a satisfying workplace almost always comes down to colleagues and working with others. (In my current office there are not one but two people with world-class laughs! Wish I could bottle it up and take it home with me!)

Let me illustrate the "ah-ha" insight with a story. In 2015, I had the opportunity to share some of our research with the international press. A core focus of my team's research is to study how people negotiate interactions on the road, for example, stopping at intersections, crossing crosswalks, or merging lanes. We have been trying to anticipate what life will be like moving through public spaces when driverless cars are on the road. Knowing that experiences vary and even simple ways of seeing the world take on different meanings, we use ethnographically inspired techniques, focused on people's actual experiences, interpretations, and understandings. So to anticipate how people will interact with autonomous vehicles (AV)—not as drivers, but as other road users—we start by examining what happens on the road today to analyze how the subtle, interactional space of roadway encounters happen and feel.

I was describing this story to the press, showing video of real street scenes from the around the world and anticipating how we might provide solutions aimed to assist in a future driving ecosystem replete with driverless vehicles. This is something few people have ever thought of. And the energy in the room was palpable. The crowning moment came when a journalist from South Korea made that "mind blowing" gesture, bouncing his hand off his head with open fists, complete with the "*ppshh!*" sound effect!

During my education, I never would have expected I'd be working for a corporation or helping build cars. But it is clear that opportunities to open minds and inform people's understanding of the world can happen in all guises. It has been moments like this that make it all worthwhile. Driving *is* a social act, and road use *is* a vital (if commonplace) part of people's lives. Bringing our social perspective to bear on the development of AV has been deeply rewarding.

6

Françoise Colpron

President
Valeo North America

Françoise Colpron is a strategic global automotive executive with 25 years of international experience in North and South America, Europe, and Asia. She joined Valeo, a $22.8 billion tech company enabling autonomous driving, electrification, and digital mobility in the global automotive industry, more than 20 years ago. Today, she serves as president of Valeo North America (NA), a more than $4 billion business with 18,000 team members in the USA, Mexico, and Canada.

Appointed to her role in 2008, Françoise successfully led Valeo NA during the economic downturn and steep recovery, quadrupling its sales and turning around the profitability of the business.

Françoise was appointed in May 2019 to the board of directors of Sealed Air Corp. (NYSE: SEE), an innovative packaging solutions company headquartered in the United States, with $4.7 billion in sales and more than 15,000 employees serving customers in 123 countries. Françoise was appointed in July 2017 to the board of directors of Alstom, a publicly listed French company and world leader in integrated rail transportation systems, with more than €8 billion in sales. In addition, Françoise serves on the board of directors of the Original Equipment Suppliers Association (OESA), as well as the Motor and Equipment Manufacturers Association (MEMA).

In 2015, Françoise was inducted into the French Legion of Honor, the highest French order of merit. *Automotive News* recognized her twice as one of the "100 Leading Women in the North American Auto Industry." In addition, *Crain's Detroit Business* named Françoise one of the "100 Most Influential Women in Michigan" in 2016.

Françoise holds a law degree from the University of Montreal in Montreal, Canada. She is fluent in English, French, and Spanish.

Education and Lifelong Learning

How did you learn emotional intelligence and how has this helped you be successful?

Emotional intelligence is a critical skill to navigate an organization and be successful in today's global and complex work environment such as thriving in an international, matrix organization. In my opinion, it is not something that you learn through textbooks but rather something that you develop through your life experiences, throughout your life and career. With time, you learn to better recognize, understand, and control your own emotions, as well as to interpret and influence other people's emotions. Some people are born with more empathy, but it is something that we can all develop by being more aware and mindful, and by learning to actively listen.

I draw from a multicultural background. I grew up half French-Canadian and half-Peruvian and am married now to an American. I have had several expatriations and international assignments, including in South America, Europe, Asia, and Africa. This has helped me develop my observation skills and decode emotions, as I lived and worked in very different environments and cultures, sometimes in places where I did not understand the language. Emotional intelligence is important to be successful in leading global teams, given the diversity of cultural differences and personal backgrounds.

Work-Life Integration

How have you dealt with work-life integration in your career?

When it comes to work-life integration, every story is so personal. There is certainly no magic formula or one-size-fits-all. In my case, I am grateful that my spouse and I share the same values and priorities, when it comes to family and work. We both have demanding careers and need to travel for work. We try to coordinate our schedules, as much as possible, so that one of us is there for our young teenage daughter. We may not be at every one of her activities, but we make sure that one of us is there for important events—whether her performances, parents day at school, etc.

Outside of work, we spend quality time with our family and include our daughter in our activities, whenever she wants to tag along—whether going to the opera, to the auto show, or to a charity event. You need very good logistics and support, whether from family or outside help, if you do not have family in town. Although I do like the idea of fluidity in the work-life integration concept, you need to recognize that it may be challenging at times to find the right balance between work-family-community and one's self.

Ultimately, your time is limited and you cannot do it all so you need to prioritize based on what is important to you, and that may change overtime. You may prioritize your career as you start out. For example, I postponed my honeymoon because of work—looking back, I would probably handle it differently today. You may decide to turn down an opportunity or take a step back for family reasons. I recently turned down a promotion abroad for family reasons. Making those calls doesn't mean that you stop advancing in your career. Often when you close a door another one opens.

Mentor and Sponsor Relationships

How important have mentors and/or sponsors been in your own career? Have they been men or women?

Since most of my career has been in the automotive industry, most of my mentors and sponsors within my company have been men. However, I have built over the years very meaningful relationships outside of my company—through various industry organizations and networking groups—and have had many informal mentors and sponsors, men and women, outside of work. In some cases, my mentor (advisor) was also my sponsor (advocate), notably when I transitioned from a position in the Legal Department to a general management position. In order to grow from being an expert to being a leader, I had to rely on the strong support not only of my mentors but also of my team. I also believe that reverse mentoring is very important to continue to grow and remain relevant, in this fast-paced, ever-changing world. Finally, I cannot stress enough the importance of networking and building meaningful relationships, early on and throughout your career.

Taking Charge of Your Career Path

What intentional decisions have you made about your career and were there opportunities you received that you had never considered?

I made the intentional decision to take assignments abroad in the earlier part of my career, transitioning from Montreal to Hong Kong for two years and then to Paris for five years. This allowed me to expand my scope, and it was a fantastic experience, both on professional and personal levels.

I later received the opportunity to transition from a function (general counsel) to a general management position (president of the region) that I had never considered. I took this opportunity, although it meant leaving my legal expertise and comfort zone, as it was a great way to broaden my scope and responsibilities. This allowed me to learn more about the business and have more exposure to stakeholders outside of the company, thus expanding my network. In both cases, I took considerable risks, but it is common knowledge that you grow more when you push yourself outside of your comfort zone.

In order to continue to grow and expand my scope, I am now pursuing board opportunities. I am currently serving on the corporate boards of Sealed Air Corp. and Alstom.

Finally, in order to share and give back to the community, I serve on nonprofit boards such as the Motor & Equipment Manufacturers Association (MEMA) and the Original Equipment Suppliers Association (OESA). On a personal level, I joined the board of the Michigan Opera Theatre, a very fulfilling opportunity.

Resilience

The automotive industry requires resilience especially with the cyclical nature and demands—what gives you the internal fortitude to keep going?

Today is an exciting time to be in the automotive industry, with the transformation of our business in terms of autonomy, electrification, and digital mobility. Valeo is extremely well positioned to capture these growth opportunities. The pace of change is breathtaking, and in order to be successful, we need to recruit and retain the best talent. I am very proud of our team in North America. Ultimately, success is all about people. It is the people I work with, at Valeo and in our industry, that give me the fortitude and the energy to keep going in the automotive industry. It is the resilience of each one of our team members that keeps the team going.

Personal Satisfaction

What would you say gives you the most satisfaction in your career?

Being people oriented, what gives me the most satisfaction, is building meaningful relationships and making a positive impact. My team gives me the most satisfaction. I enjoy leading people as opposed to leading tasks…and winning both their hearts and their minds.

I also enjoy building meaningful relationships with all of our stakeholders, including the communities in which we do business. Beyond making a positive contribution at work, I strive to give back. I am particularly interested in mentoring our girls to become strong, independent women who have equal access to opportunities and the skills to succeed—women who have the choice of having a successful career while raising a family or pursuing other personal goals.

Alicia Boler Davis

Vice President, Global Customer Fulfillment
Amazon

Alicia Boler Davis is vice president, global customer fulfillment at Amazon. She joined Amazon Operations in April 2019 and has responsibility for the worldwide network of over 175 fulfillment centers across 16 countries.

Prior to joining Amazon, Alicia was executive vice president, global manufacturing for General Motors – a role she held from June 2016. Boler Davis began her GM career in 1994 as a manufacturing engineer at the midsize/luxury car division. During her GM career, she held many positions of increasing responsibility in manufacturing, engineering and product development, including plant manager of the Michigan Orion Assembly and Pontiac Stamping facilities, as well as vehicle line director and vehicle chief engineer, North America Small Cars and leader of GM's Global Connected Customer Experience. Under her leadership, GM improved vehicle quality and fundamentally redefined customer care and its interaction with customers through social media channels and customer engagement centers.

Numerous organizations and publications have recognized Boler Davis for her professional accomplishments and community service, especially in the area of science, technology, engineering, and mathematics (STEM). Of note, in 2018 she was named Black Engineer of the Year by Career Communications Group and one of the most powerful female engineers by Business Insider. Also, in 2018, she received an honorary doctorate in engineering from Rensselaer Polytechnic Institute.

She is a member of the Northwestern University McCormick Advisory Council and a member of the General Mills board of directors. Boler Davis has a bachelor's degree in chemical engineering from Northwestern University, a master's degree in engineering science from Rensselaer Polytechnic Institute and an MBA from Indiana University.

Education and Lifelong Learning

How have you structured your own approach to lifelong learning?

My commitment to lifelong learning is very important to my ongoing development as a leader, colleague, daughter, wife, and mother. I like to set ambitious goals, strive for excellence, and enable my teams to do great work, and a big part of my continued development is a willingness to embrace learning each day.

As an engineer, I am inquisitive by nature. I believe a curious mind is a receptacle for learning. Even from an early age, I was very curious about how things work. I would fix appliances that broke around our household and would learn how things worked by reading and asking questions.

Throughout my life, I have had an ongoing curiosity for knowledge for either personal or professional growth. I learn by reading, experiencing, and listening. I think listening is a very important component to learning which includes listening to audio-books, podcasts, and, more importantly, truly listening to what people are saying and not saying. In addition, embracing feedback on your performance and behavior is also key to learning. Do not take criticism personally, as it may help you overcome your shortcomings and can be a way of learning from your mistakes. Continuous improvement and lifelong learning should be your goal.

Learning is a priority in my life, but it does not happen by accident. Nothing will happen unless you make it happen and put in the effort.

Work-Life Integration

How have you dealt with work-life integration in your own career?

With the hustle and bustle of everyday life, the keys to work-life integration are having a clear set of priorities, being very organized, and having a schedule. In addition, making that integration a priority is a critical element as well.

While I have a tremendous amount of responsibility with my work, I am also very proud to be a mom of two teenage boys, who are both very involved in school and athletic activities. My husband and I share the responsibilities to make sure we are both engaged in our children and each other's lives. We have our struggles at times, just like other busy, professional couples, but being actively engaged in our children's lives and work-life integration are very important to us, and we work hard at it to keep it a priority in our daily lives.

I think it is also very important to realize that you cannot do everything by yourself and, at times, you need help. Family is very important for those times where a helping hand is critical. When my kids were younger, my mom played a big role in helping keep everything running on time. Without the support of our family, we would certainly struggle more than we do.

Mentor and Sponsor Relationships

How important have mentors and/or sponsors been in your own career? Have they been men or women?

Mentors have played a very significant role in my career. While there have been several, I specifically want to acknowledge and highlight Bill Boggs, a retired former GM manufacturing and quality senior leader. He was very instrumental in the development of my career. I met Bill when I was a young employee working at GM's Detroit/Hamtramck assembly plant. He saw something in me that I did not necessarily see in myself. I am forever grateful for that. On many occasions, he pushed me out of my comfort zone and challenged me to take on assignments that were more significant.

Early on, he told me that I had the potential to be a plant manager. His assessment was surprising, but with his support and nurturing, my confidence grew in my abilities. While I had my share of failures, he helped me learn quickly and move forward. Throughout my career, he was there to provide input, suggestions, and guidance and share his experiences. Today, I remain close friends with Bill and his wife, Nancy.

I have personal experience on how important giving back is to the performance of your organization, but it enhances your ability to lead. I get tremendous personal enjoyment and fulfillment out of mentoring others. I feel obligated to give back to others, because so much was given to me. I want to help those coming up in the organization behind me.

Resilience

What qualities make up resilience in a leader?

I pride myself on being a resilient leader. I consider a resilient leader a leader who sees failures as an opportunity in disguise. A resilient leader is one that can learn and recover from failure quickly. As an engineer, I understand and embrace the reality that life is full of challenges and opportunities. A resilient leader embraces challenges and uncertainty. I think it is critical to remain positive, open, flexible, and willing to adapt to change.

My parents divorced at an early age, so I learned early on that life would be filled with challenges. Not everything will go your way or as planned, and the way you approach challenges and problems will help provide the foundation for your success.

Whether in my professional or personal life, I am very optimistic. I believe everything happens for a reason. I also believe you get back what you put out. I treat people the way I want to be treated. I think consistent leadership is important to the success of any team. There will be difficulties in any organization, especially with the cyclical nature of the auto industry, but consistent, unwavering leadership is very important. Resilient leaders maintain a positive attitude and a strong sense of opportunity during periods of difficulty and turbulence.

In the end, regardless of the issue, I think attitude plays a huge role in the success of an individual or team, during the good times or the tough times. I think it is important to show up every day, get in the game, and be the example for others to follow.

8

Corinne Diemert

Vice President of Sales
Valeo North America

Corinne Diemert joined Valeo, a $22.8 billion tech company enabling autonomous driving, electrification, and digital mobility in the global automotive industry, nearly 30 years ago. Today, she serves as vice president of sales for Valeo North America, a more than $4 billion business with 18,000 team members in the USA, Mexico, and Canada.

Corinne leads the sales growth of Valeo in North America, as well as all customer-focused sales initiatives with General Motors worldwide. She is a member of Valeo's North American leadership team and directs a diverse group of more than 70 high-performance sales professionals.

Corinne builds on a multifacted professional background, beginning her Valeo North America career in plant logistics and then transitioning to program management. Subsequently, she was appointed to leader of operations launching the company's first manufacturing site in the Greater Detroit area—Valeo's Highland Park front-end module facility. In her current role as vice president of sales, Corinne's leadership contributed to the company tripling its sales over the past five years.

Corinne is the U.S. Champion of Valeo Women Connected, a worldwide network devoted to the career development of women in the organization. Additionally, she is on Inforum's AutomotiveNext executive committee and a member of the Original Equipment Suppliers Association's sales executive council.

Corinne holds a degree in business administration-material management from Conestoga College in Guelph, Ontario (Canada).

Education and Lifelong Learning

How did you learn emotional intelligence and how has this helped you to be successful?

Early on in a career, young team members are like sponges, often building a professional personality based on how mentors are responding to situations. Sometimes they pick up bad habits in the process. At a crucial point at the beginning of my career, I realized my behaviors were reactive to negative situations rather than aligned with positive leadership styles or even with my own authentic self.

With experience comes self-confidence and my ability to increase my self-awareness to reflect on my strengths and values to understand what was important to me. As I progressed in my career, I overcame my initial fear of criticism and learned from the reaction of others how to find the right balance of being positive and drive excellent outcomes—not only learning how to manage my own emotions but how to read and be more empathetic to those around me. This helped me to build a character that allowed me to rise in my professional environment as a leader, a mentor, a teammate, and a supplier.

Work-Life Integration

How have you dealt with work-life integration in your own career?

I started my adult life believing that you can have it all—a great career, family, friends, and happiness in my personal life. I still hold this belief, but throughout my career, there have been times when certain of these elements have become imbalanced. The important thing is to recognize when things get out of balance and to make the necessary adjustments.

Making corrections to find balance required me to focus more on quality of time, not quantity of time. I had to learn how to step away and detach myself from work stresses and to be truly present when I was at home with family and friends. I needed to learn to say "no" and to realize that I could not be everywhere. Finally, I realized the benefit of being organized and to delegate, both professionally and personally, to optimally utilize the time available.

If you do not manage your balance you may feel cheated by your employer or by your family. Ultimately, a person who finds the right balance is happier and more productive. This is one of the reasons I am a big supporter of flexibility in the workplace, both personally and with my team.

Mentor and Sponsor Relationships

Is it important to have a structured system in a company for mentoring or should it happen synergistically or both?

It is absolutely critical to have informal mentoring opportunities in the workplace, and if you can combine it with a formal mentoring program, it is even better. An informal mentoring program starts simply with a voluntary relationship or friendship that helps team members develop and learn from each other. This requires leaders to develop an open and positive work environment where team members trust and demonstrate appreciation for one another.

If you are working in a company that does not offer a formal mentoring program, don't let this prevent you from realizing the benefits. You can always create your own informal mentorships. I have learned that some of the relationships I have built through informal mentoring have not only enhanced my career but, in some cases, resulted in lifelong friendships.

At Valeo, I have also had the opportunity to participate as a mentor in a formal mentorship program and I have realized the benefits for both the mentor as well as the mentee. I was able to learn about other departments within the organization and increased my ability to listen to some of the challenges our team members face. Thereby I improved my personal leadership skills with my own team.

Taking Charge of Your Career Path

What is the biggest surprise in your career when you faced a new opportunity thinking what was a terrible move, but tackling it in any case—what did you learn and how did it change your view?

In 2008, I transferred to a position in operations responsible for launching a new production facility in the Detroit area during the economic downturn. At the time, we faced considerable uncertainty about the future of our industry in Detroit, the vehicle launch, and our program. At the time, many of my colleagues questioned my decision of taking on this challenge that in their eyes was destined to fail.

From this experience, I learned how to build and depend on a strong team focused on collective objectives within our control and how to avoid stress and distraction from the conditions that were outside of the team's control. The result was the most rewarding job experience of my career. Our team pulled together to ensure success in an environment with common goals and objectives, never politics. We built lasting relationships within our teams and customers, based on trust and transparency.

Do not shy away from a challenge. Do not automatically take the safe route. You may miss out on the best professional learning experiences of your career.

Resilience

What are the biggest drains on resilience in the workplace?

I believe office politics, negative attitudes, and bureaucracy are the biggest drains on resilience in the workplace. As leaders, we need to focus on teamwork and trust to build a nurturing, positive work environment. This starts at the top. It is how we, as leaders, interact and work as a team that sets the stage for the broader work environment. Our assessment of our team members must not only assess results, but how they achieved those results.

Personal Satisfaction

What would you say gives you the most satisfaction in your career?

In order for me to be happy at work, I need the ability to make a difference.

You may wonder how I continue to stay motivated and achieve this satisfaction after working at the same company for nearly 30 years. Valeo has offered me the opportunity to continue to add value and to challenge myself throughout my career.

As part of Valeo's leadership team, I receive the most satisfaction from professional coaching opportunities and leading a team to achieve success.

Lisa Drake

Vice President
Global Powertrain and Purchasing Operations
Ford Motor Company

Lisa Drake is the vice president, global powertrain and purchasing operations, serving in this role since January 2018.

In this role, Drake is responsible for all powertrain purchasing operations worldwide and global operational purchasing performance. She reports to Hau Thai-Tang, executive vice president of product development and purchasing. She joined purchasing in 2013 as director of global program purchasing, and most recently served as director, global interior purchasing, a position she held since August 2016.

Before purchasing, Drake held various positions in product development for nearly 20 years. In 2004, she led the F-150 product and launch team during the construction of the state-of-the-art Dearborn Truck Plant at the Ford Rouge Center. Her other notable positions include chief engineer, Lincoln MKC, assistant chief engineer, F-series Super Duty, and program manager roles for Explorer and Expedition.

Drake also served as the global hybrid/battery electric vehicle chief engineer from May 2007 to September 2010. In this position, she led the development and delivery of Fusion hybrid, Lincoln MKZ hybrid, and the C-Max hybrid and Energi programs.

In addition to her work with hybrid and plug-in hybrids, Drake led the development of the Focus Electric and the Transit Connect Electric. In 2008, the Automotive Hall of Fame awarded her the Young Leadership and Excellence Award, in recognition of her contributions and leadership in the growing field of electrification. Drake joined Ford in 1994 as a Ford College Graduate in powertrain engineering. She holds a bachelor of science in mechanical engineering from Carnegie Mellon University and a master of business administration (MBA) from the University of Michigan.

CHAPTER 24

Education and Lifelong Learning

How important is an MBA or other graduate degree?

Certainly, it's hard to go wrong in getting an MBA or other advanced degree. But I hesitate to say it is necessary for success, as I know several leaders who have various types of educational backgrounds, some of which do not include graduate work.

For me, I always knew that I wanted to pursue a graduate degree after Carnegie Mellon, but like many, I wasn't sure which path at first. The MBA or the masters in engineering? I ultimately settled on the MBA mainly because I realized that I wasn't exactly using a lot of the deep university engineering theories in my daily work, so getting a masters degree with even more depth didn't seem like it would help me. Plus, I was becoming increasingly interested in the bigger picture of the Ford business, so an MBA seemed like a more natural progression.

I started my first semester in business school at the University of Michigan campus in Dearborn two years into my time at Ford. This was very convenient given that it was less than three miles from my office. I, along with eight or so other Ford engineers, would essentially cross the street, two nights a week. The work was demanding, but it never became very fulfilling. With all of the Ford employees in the class, plus others also largely from the auto industry in the immediate area, it felt a bit like an extension of the workday. So a few of us made an agreement that we would sign up for the main campus classes the next semester and make the trek to Ann Arbor in the hope of finding some variety. This turned out to be a great decision, as our classes were no longer dominated by auto engineers just like us. In our courses we had burgeoning CEOs, doctors, and medical professionals looking to advance to management positions, economists, and so many other interested learners, of all ages, with such unique backgrounds. It was the interaction with those bright and curious people that gave my MBA experience the greatest value.

Many early career employees come to see me asking about the value of a graduate degree and in which way to pursue one. I explain that the coursework is only 50% of the learning. The collaboration, the exploration, and the self-awareness that you develop when you have to learn to deliver heavy project-work with people of vastly different backgrounds is the other 50%. It's not just about checking the box with the degree. Then I ask them to be very wary of the short view with online class work. Take the long view, put in the work while you have the time to do it, and really maximize the value of that time with people who challenge you in a way in which you will indeed be challenged again at some point in your career. It will help you to be ready.

Work-Life Integration

What do you think about the Lean In concept?

I didn't think much about it, until…

Every year, Ford holds a large employee event in our world headquarters auditorium to celebrate International Women's Day. A few years ago as part of the program, we had the honor of having Sheryl Sandberg speak to a standing-room-only full house of women (and some men) right around the time of her book publication. I was running a few minutes late from a meeting prior, and had slipped in a door at the back of the auditorium and found an open space against the back wall. Sheryl was on stage talking about her experiences, about leaning in and not discounting your capabilities, and to be honest, I didn't relate. I had managed to be successful through most of my career by just working smart and not thinking about all of that. Then she came to a point in the speech where she was talking about how assertive women are perceived differently than assertive men—the bossy part of the concept. She described how little girls are labeled bossy so early in life, differently than boys who are labeled as leaders when exhibiting the same behaviors. And then she asked the audience, "How many of you have ever been told you are too aggressive at work?" I thought I would see a few brave souls raise their hands—because let's admit it, it happens. What I witnessed, though, shocked me. From my vantage in the back, I saw a thick wave of hands go up across the entire audience. A lot of head nods. Even some tears.

I felt my heart drop into my stomach. This struggle that I thought was likely affecting a few women in my workplace was actually affecting hundreds, if not more.

It occurred to me in that moment that I hadn't really embraced the Lean In concept because it wasn't written for the minority of women like me who had somehow persevered. But for every one success story, there were hundreds of women that weren't being seen for their value, and were struggling with what tools to use and how to succeed. Lean In was relevant to them. And that's when I realized that my part in the Lean In movement was to be an advocate and a sharer to help other women be more successful in achieving their aspirations, whatever those might be, which is why I also agreed to write this chapter.

Mentor and Sponsorship Relationships

How important have mentors and/or sponsors been in your own career? Have they been men or women?

I can't even imagine how anyone can be successful in building a career in a large company without having a mentor and, just as important, an advocate. Sometimes those are two distinct roles—a mentor can help with career guidance and questions. An advocate knows your work, and will personally advocate for you in your career.

And through most of my early and mid-career, my mentors and advocates were predominantly men. Not for any other reason than there just weren't as many women in leadership positions to engage in mentoring. Thinking back, the gender of my mentors never really crossed my mind. They were just great human beings who genuinely gave honest and relevant guidance. I would not be at the place in my career today without them. I never had a set cadence with any of my mentors, and our conversations were more organic in nature—conversations around strategic business imperatives or career moves I was thinking of making.

I often confess to people who ask about the benefits of mentoring that I have this somewhat abstract concept that I call "Fantasy Ford"—it's my version of "Fantasy Football" but with a "team or bench" of advocates, colleagues, and leaders that I can rely on for various things such as career advice, ways to check my strategic thinking, connections to get things done. You have to go out and consciously build your bench, develop your connections, and actively share your ideas and strategies. Anything worth changing in large companies needs a lot of teamwork—you can't go it alone. So I have built a bench over time and rely on it constantly as I endeavor to make change. And the higher I advanced in my career, the more expertise I found I needed to add to my bench.

And to be clear, this bench building has to be done with the right intentions and with a genuine interest of learning from these individuals and absorbing how they see the business from a different angle. It's not about self-promotion. You do have to demonstrate your accomplishments and your capabilities, but the time spent building your bench is around understanding the views and perspective of others and using that to connect strategies to advance the change you want to make.

As a final thought, although I would say that my formal mentors, those with whom I would schedule office meetings for insight and counsel, were men, my network of female peers is what buoyed me my entire career. We had a group of us in the late '90s when I was in the Light Truck Division who were brought together (by a male vice president advocate) and we were called the "Women Of Truck", WOT for short. It was an extraordinary group of successful women in a very male-dominated truck world, and we built networks and friendships that have endured. Some have retired, some have moved on to other careers, and some are still at Ford. They may have no idea how important they were and are to me until they read this. I hope I have been able to be there for them in reciprocity. If you haven't seen Abby Wambach's Barnard College commencement address, it's time well spent. Abby calls networks like this "wolf packs." Wolf packs, "Women Of" groups—call them what you want. Create your network and be confident about it, and rely on it.

Taking Charge of Your Career Path

What intentional decisions have you made about your career and were there opportunities you received that you had never considered?

I think my career is almost an equal mix of intentional choices about what I wanted to do and what skill sets I wanted to develop, and surprising twists in my career path that I would never have anticipated.

When I started at Ford, I wanted to challenge myself with engineering the hardest systems first. I was used to an intense academic environment and wanted to continue to prove myself and set the bar as high as it could go in my new job. I had several offers from different divisions at Ford but chose powertrain engineering in the truck division because I knew it would be an enormous stretch for me. And it was. But as I was finishing my MBA, I really wanted to move into an area where I saw more of the overall business. Through great mentorship, I orchestrated a move into program management of vehicle projects. I spent several years in various roles, being promoted a few times through those assignments. But I started to feel that I was becoming too removed from the product engineering work, so asked for an assignment as a plant vehicle team (PVT) manager which is essentially the product engineering leader of a small team that was resident at an assembly plant. As luck would have it, Ford's Dearborn Truck Plant was under construction at the time, and I was able to transition into the PVT manager role for that plant and be a part of the operating committee. And I felt good that I self-identified an area where I wanted to develop, found an opportunity, and filled some of the technical gap just as I had intended. And I had a renewed appreciation for Ford in that I didn't have to move companies to gain critical experiences that I felt I needed. Ford had options.

But there were also two very distinct career path changes that I never saw coming. And both of them were probably the most pivotal in my career.

The first was my move from the assistant chief engineer on the Super Duty to the hybrid/battery electric chief engineer. Moving from one of the largest trucks in our lineup (think big diesel, high-torque engines) to vehicles with no engines, batteries, and power electronics that required me to pull out my electrical engineering (EE) books from college was a radical change, to say the least. I never thought I would leave truck, but a promotional opportunity was presented to me, and I decided to take a major leap. And as you can imagine, with today's environment of EV and AV technology advancing at never-before-seen pace, having the background from that uncharted move ten years ago has now given me a unique advantage as a procurement professional around our battery cells and EV powertrain component sourcing strategies.

The second was my move from product development to purchasing in 2013. Before that time, it wasn't very often that individuals moved between the two organizations in Ford, especially at certain leadership levels. This again was not a move that I originally had on my career plan. But one of my mentors who, in hindsight I can now see, knew my capabilities more than I knew them myself, suggested that I take a position in purchasing. I followed his advice, and I can say that the new environment and new learning curve allowed me, or perhaps required me, to spread my wings so much further than I thought I could. I was visible to more of the executive leadership, and my agility to assimilate and then lead after being dropped into a foreign space was noticed. As a result, my career trajectory changed dramatically after that move.

Through personal experience, I now have a much deeper appreciation for the potential that is unearthed in employees when they take more unorthodox steps in their careers. In mentoring conversations, I try and encourage individuals to think about risk and reward and to find places in their career development planning where they think it can fit. The safe path is just that.

Resilience

Do you consider yourself resilient? If so, how did you become that way?

I was fortunate to grow up with great parents and a stable household by most measures. And I can recall at an early age hearing my mother tell my grandparents or aunts and uncles how well I was doing in school. I'm not even sure she ever knew I was listening to these conversations, but I heard her. And that reinforcement, coupled with watching her tackle anything she wanted to do, fearlessly, is where I think my self-confidence took root very early. My father is one of the most social people I know—he knows everyone, and everyone knows him. He can walk into a restaurant almost anywhere near my hometown, and it is inevitable that he strikes up a conversation with someone. He has incredible social confidence. Confidence is just something they both had, and assuming one is partly a product of upbringing and environment, it's not surprising that I became a confident individual. And for me, it's my confidence that supports my resiliency—I always believe things can be fixed, or that I can change direction if I find myself at an impasse. There isn't much that seems to be impossible to do in my mind, and failure is okay if you dissect it, learn from it, and make another attempt.

Also, an ability to look at things in a relative way is critically important so that you can put things in perspective before the emotions take over, you can save the resiliency for the issues that really matter, because a lot of what we stress over doesn't really matter.

Personal Satisfaction

What would you say gives you the most satisfaction in your career?

There is a TED Talk that Bill Ford gave in 2011 that you can view on YouTube. I think I have watched that talk at least 10 times and I show it often. I use it every year when I host recruiting events for Carnegie Mellon, I show it as part of my self-introduction when I join new teams at Ford, and I mention it at times when I speak externally at events. Bill connects the ability to move to the welfare of our society. Movement gives people the opportunity to get to health care, to get to jobs, to unleash the gridlock that prevents progress. When I watch it, I am reminded at the highest level why I really love to work at Ford. It's because through our work, we are making people's lives better. We are making the world a better place.

When I joined purchasing in 2013, it was an incredible realization for me how large of a socioeconomic impact can be made with a portion of Ford's billions of dollars of annual purchasing power. This hit me square in the heart the first time I visited Detroit Manufacturing Systems.

I was a new director in purchasing, in charge of the global interior commodity buy—all of the seats, door trims, headliners, restraints, and associated interior parts in our vehicles. I had been given my minority spend objectives when I started. Essentially a certain percentage of the sourcing was targeted for minority, veteran, and women-owned businesses. Of course, as with any objective, I intended to meet it, and with some legwork with our supplier diversity development team was able to put a plan on paper in short order.

About 4 months into my assignment, I had a trip planned to one of our largest interior supplier manufacturing sites—Detroit Manufacturing Systems. Andra Rush was the CEO, and I was going to meet her at the facility for a plant tour. I knew from others that Andra was a passionate advocate for helping serve underdeveloped areas of the community, and DMS was one of her larger projects. I drove up the Southfield Freeway from Dearborn and exited the highway traveling through an area that had seen tough times. All of a sudden this shiny new plant comes into the horizon. As I pull up to the gates, I'm greeted by the most wonderful older gentleman, and he checks me into their system. Took a little while, but he got it done. Inside the building I am greeted by a staff as diverse as you can imagine. And as we are traveling through the plant floor on the tour, people are coming up to Andra thanking her for supporting their jobs, allowing them steady incomes, and affording them the ability to make their own money. The factory was buzzing and full of hardworking men and women who might otherwise not have had the gainful and safe employment and ability to make a better life for themselves and their families had it not been for this plant, and for Ford's commitment to invest in and develop the areas in our community that need it the most. That's when I became more fully aware that in our quest to build cars, trucks, and mobility solutions to give people the freedom of movement as the endgame, we are actually helping people all along that way. Being a part of *making people's lives better* is what gives me the most satisfaction.

10

Bonnie Van Etten

Vice President, Group Chief Accounting Officer
Fiat Chrysler Automobiles N.V.

Bonnie Van Etten is the group chief accounting officer for Fiat Chrysler Automobiles N.V. (FCA) and is responsible for overseeing the company's accounting, reporting, and internal controls. She was named to the position in March 2017 and is based in Auburn Hills, Michigan.

Prior to her current role, she was the vice president-chief accounting officer for NAFTA and FCA US LLC.

Van Etten has handled a number of increasing financial responsibilities within the company including serving as head of global technical accounting and accounting research for Fiat Chrysler Automobiles N.V. and FCA US.

She joined the company in December 2010 from American Express, where she last served as vice president for regulatory reporting. She received her bachelor's degree in finance from Anderson University, Indiana (USA).

Her work background includes

- 2017: Current, vice president, group chief accounting officer-Fiat Chrysler Automobiles N.V.

- 2015: vice president, FCA-NAFTA, FCA US chief accounting officer

- 2013: head of global technical accounting and accounting research, Fiat Chrysler Automobiles N.V. and FCA US LLC
- 2010: director, technical accounting—FCA US LLC
- 2009: vice president, regulatory reporting, American Express Company
- 2006: vice president, controller—technical accounting group, American Express International, Inc.
- 2001: director—capital markets group (Europe), PricewaterhouseCoopers LLP
- 1997: U.S. audit practice, PricewaterhouseCoopers LLP

Education and Lifelong Learning

How have you structured your own approach to lifelong learning?

My approach to life long learning is multifaceted and dynamic. Since I was a child, I've been naturally curious and asked a lot of questions. Asking questions is a skill I've carried with me into the workplace. No matter the position or company, asking questions has been paramount because I believe it is the best way you can learn.

I've also learned a lot about myself by taking chances, such as living overseas, taking a role that wasn't my first (or second or third) choice, and moving to a new country in a new job in a new company with no safety net. This is not only applicable to my professional life but also my personal life. I've had many wonderful experiences by taking chances and pushing myself which has led me to learn to scuba dive, finished an Olympic distance triathlon, and drive a race car.

In addition, some of the most valuable lessons I've learned have been when things did not go well or turn out as planned. For example, when I took a position that required creating a brand-new team to manage new reporting requirements, I made several leadership mistakes in how I managed the team. I've learned from those mistakes and have had the opportunity to, when encountered with a similar situation, take a different path that led to a more positive outcome. In conclusion, I believe that if we do not strive to learn and grow every day, we risk becoming stagnant and irrelevant.

Work-Life Integration

Is work-life integration more possible at some career levels than others?

To me, work-life integration is a very personal concept as it requires an individual assessment of what one considers to be important, what one is willing to sacrifice (or not), and what consequences one is willing to accept. I also believe that work-life integration is a journey as there are points in time when it requires a different balance than at other points in time. With that being said, based on my personal experience, at the higher levels, there is often a greater level of flexibility that may exist at lower levels. At the higher levels, the expectations are higher and, usually, the demands are greater; however, I think in some companies there is more flexibility given on how to juggle those demands and expectations. For me personally, I recognize the fact that my employees often have other priorities than just work and try to provide as much flexibility as possible for them to balance work and life. I also recognize that my employees are more engaged and productive when they are able to integrate their work and lives successfully.

Mentor and Sponsorship Relationships

How important have mentors and/or sponsors been in your own career? Have they been men or women?

Mentors and sponsors have been instrumental in my career. I would not be where I am today without the help of mentors who were always there during some of the most critical points of my career. They pushed me to stretch and accomplish what I thought were tasks that couldn't be achieved. In one particular instance, the partner I was working for was unable to be on-site during a very critical time for a client.

This partner had the confidence that I could lead this part of the project successfully and manage the complex issues to resolution. His confidence in my abilities was key in helping me to believe in myself. The project was successful and I came away with a new found confidence in my ability to handle more than I previously thought I could. These sponsors also presented me with opportunities which helped me grow both personally and professionally. I've had both men and women sponsors throughout my career; however, I can only recall two women sponsors as opposed to more than five men sponsors. During the early years of my career, I didn't recognize that I had sponsors, and it was only in reflection that I realized how lucky I was to have had them and how they positively influenced the direction of my career. I learned something different from each of my sponsors, and they have all had a hand in helping me to become the leader I am today.

Taking Charge of Your Career Path

If you changed companies, what was the compelling reason and was the move beneficial? Why?

I have changed companies twice in my career (so far at least!). I was compelled to change the first time by an overwhelming desire to move out of public accounting and into the industry where I thought I would have better work-life balance and be able to manage my working life in a more constructive way. Working with clients was not always my strength and I found it hard to say "no" or push back against unrealistic timelines. Little did I know at the time but the skill of being able to say "no" or push back against unrealistic timelines is not unique to public accounting.

While I did have many adventures in my personal life, I also grew in my professional life and it challenged me in ways that were unexpected. The second job change—my move to Fiat Chrysler Automobiles—was made for a combination of reasons. I primarily wanted to be closer to my family, but I wanted to work for a company that made something tangible. I also wanted to work at a place that provided me with an opportunity to grow my career. My arrival coincided with the transformation of Chrysler to Fiat Chrysler. It is rare to experience such a journey.

Personal Satisfaction

What gives you joy in your job? What causes you the most angst?

That moment when I see the light bulb click on for people is one of many joyful moments I experience at work. I also find joy in mentoring and helping others connect to enable them to reach their goals. I enjoy helping my employees and others to move into different roles that help them continue to learn and grow, even if they have to leave my team to do so. The angst moments are delivering tough messages even if it is the right thing to do and the inability to promote high potential individuals because the right job either doesn't exist or isn't available. Lastly, I never like asking people to give up their family time to meet deadlines.

Joy Falotico

President, Lincoln Motor Company and Ford Chief Marketing Officer

Joy Falotico is president, Lincoln Motor Company and Ford chief marketing officer. In this role, Falotico is responsible for leading the continued evolution of Lincoln as a world-class luxury brand and oversees all Lincoln operations globally.

Falotico also leads the company's marketing function and efforts to connect more closely with customers by identifying new opportunities to serve them. She now reports to Joe Hinrichs, president, Automotive.

She is also chairman of the Ford Motor Credit Co. LLC board of directors.

Previously, Falotico was chairman and chief executive officer of Ford Motor Credit Co., a leading global financial services business that supports Ford dealers and customers, and the sale of Ford and Lincoln vehicles. Prior to that, she was chief operating officer, leading Ford Credit's global operations in the Americas, Asia Pacific, Europe, and the Middle East and Africa, as well as marketing, sales and brand, business center operations, and insurance operations.

Falotico also was executive vice president of Ford Credit marketing, sales, Americas, and strategic planning where she had the responsibility for marketing and sales globally, and business operations in North and South America.

Since joining Ford Credit in 1989, Falotico has served in a number of senior positions, including executive vice president of Ford Credit North America.

Falotico is involved in the American Financial Services Association (AFSA), a U.S. financing industry trade organization, previously serving on its board of directors and executive committee, and as chair of the AFSA vehicle finance division board.

Falotico holds a bachelor's degree in business administration from Truman State University and a master's degree in business finance from DePaul University.

Education and Lifelong Learning

How did you learn emotional intelligence and how has this helped you be successful?

Having the opportunity to be a people leader very young in my career allowed me to learn the importance of emotional quotient (EQ) early on. It's so important to be self-aware as a person and as a leader. I remember being shocked to find out that how I might be perceived could be very different than I thought. I was fortunate that our company had annual employee satisfaction surveys and peer and management leadership feedback processes that allowed me to gain valuable feedback on how I showed up as a leader.

Establishing my values as a leader and my personal brand which is what I stand for and what people can count on to be true working with me was very important. Actually writing it down and sharing it creates awareness for others and puts me on the record to stay true to my values. We spend a lot of our time/life in the workplace and being able to develop strong working relationships with your colleagues is important. These relationships have to be built on trust and a mutual respect. Also leaders have to lead through good times and bad, and EQ is key to successfully navigating choppy waters with your team. In my experience, people will run farther and faster for leaders they respect and can trust knowing they have the best interest of the company and team in mind when making critical decisions.

Work-Life Integration

Have you ever chosen work over family or vice versa? Has this gone well? What did you learn?

There have been circumstances when I chose work over family events but my family knows I would rather be with them which helps them to understand and/or forgive me. I do make it a point to be at the important events and to be dependable. I have never said I would do something with my family and then not done it or had to back out, and I've always been clear up front so my word is good and there are no disappointments. I feel very lucky that I have a very engaged spouse who also works but shares the duties with me so one of us can be there for our kids.

Mentor and Sponsor Relationships

How important have mentors and/or sponsors been in your own career? Have they been men or women?

Most of my mentors have been men and they have been my boss. They were more informal and were genuinely interested in providing instrumental career feedback and encouraging me to keep my mind open to opportunities that would build my business acumen. I also benefited from a formal mentor relationship that was more cross-functional in nature that allowed me to learn about areas outside of my current skill team.

Taking Charge of Your Career Path

What intentional decisions have you made about your career and were there opportunities you received that you had never considered?

I have taken on roles with the intent to grow my breadth of knowledge to support my career trajectory. It meant a heavier work load but it also meant more learning and experience gain. The role I'm in now is one that I never considered and is a great example. I have learned more from taking the leap than I would have if I stayed in my prior role which has helped to reaffirm my decision.

Resilience

The automotive industry requires resilience especially with the cyclical nature and demands—what gives you the internal fortitude to keep going?

Knowing we can win gives me internal fortitude. This is my 30th year in the industry and I've seen many cycles, and I know we have great people that in the face of adversity will come together and do the right thing. Also, knowing I can contribute gives me resolve as there is so much opportunity in this industry and so much to do. As a leader, I can usually see and map a way forward and that also gives me strength and fortitude. This doesn't mean that it's going to be easy, but I always believe we will carve a path and find a way to win.

Personal Satisfaction

What would you say gives you the most satisfaction in your career?

The fact I can come to work every day, make a contribution and a difference—some days small and some days bigger. But I derive a lot of satisfaction by delivering results with the team. I feel like the company and my team are counting on me and it's my duty to not let either of them down. That probably sounds old school, but it's the truth!

12

Pamela Fletcher

Vice President-Global Innovation and R&D Laboratories
General Motors Co.

As vice president of Global Innovation and Research & Development (R&D) Laboratories, Pamela Fletcher oversees the technology-led business transformation of General Motors.

She leads the teams at GM whose mission is to disrupt the traditional automotive industry. To her position, she brings an established track record of delivering multiple products to market, ahead of the competition and with better results. Pam has overseen the development of critical technical and organizational capabilities that position General Motors for large-scale production of electric and autonomous vehicles.

At R&D, she directs seven R&D laboratories around the globe that are focused on various future mobility technologies.

For more than a decade, she has been in leadership roles guiding the development of GM's electric vehicle and self-driving technologies, most recently as vice president, global electric and autonomous vehicles.

Fletcher earned her bachelor's and master's degrees in engineering, and serves as a corporate director of Coherent Inc., a NASDAQ-listed company based in Silicon Valley.

Fletcher was named to *Motor Trend's* 2018 and 2019 Power List of auto industry leaders and was one of *Fast Company's* "Most Creative People" of 2017.

Education and Lifelong Learning

How have you structured your own approach to lifelong learning?

I am really motivated and excited about technology and emerging technology. For example, understanding how a certain technology works and, in turn, how it can change someone's life. Originally my love of cars brought me to the automotive industry, but it has been my love of invention and new technology that has kept me learning and growing in my career. I seek assignments that involve new aspects of technology and innovation to not only keep learning myself but also to help our customers see what's possible.

Work-Life Integration

How have you dealt with work-life integration in your own career?

There isn't one perfect answer to work-life integration. An individual's needs vary throughout life, just as career needs change, too. I think the great thing is that we have a number of ways we can try to manage work-life integration, for example, communicating well and working remotely can make a huge difference.

A story that resonates most with me on this topic has to do with one of my proudest moments of my career, which was launching the first-generation Chevrolet Volt. It was such an exciting time in my career, but it was also a time that was personally difficult for my family. My father had been diagnosed with cancer and he was in the hospital right at the time when we were launching the vehicle. It was really important for me to be with him, my mom, and my family, so that's where I was. The great thing is General Motors completely supported me to be with my family, and I was still able to be engaged and to lead from afar thanks to technology. The Volt launch team did an awesome job and I was able to spend valuable time with my family.

Mentor and Sponsor Relationships

What does it take to be a good mentee?

The great thing about having a mentor, whether if it's formal or informal, is that it gives you an environment to talk about the things that matter to you. With your mentor you can bring forward questions, concerns, and also share examples of things that you've gone through recently and share how you handled them. Discussing specific situations with your mentor can help you be very introspective while gaining feedback on your approach. These types of conversations are really valuable to have.

Taking Charge of Your Career Path

What would you do differently about planning your own career if you were starting over?

From early in my career to even sometimes now, I like to be engaged in what's going on, not just operating at the highest level. I like to be able to speak frankly and intelligently about what the real opportunity is with a product or technology, discuss the real risks and how we might overcome them. Early in my career I felt that I needed to come up with all of that on my own and I didn't always put value in a network because I felt like spending time talking to others would show I didn't have the right motivation. I learned later in my career that that approach was a missed opportunity. Having a network of peers and stakeholders brings different perspectives to the table that can really enlighten you to think about the problems you're trying to solve from many angles. When you do need some help, your network is there and ready to collaborate or make a connection. I really encourage everyone to have a strong personal and professional network to help improve everyday quality of life.

Resilience

The automotive industry requires resilience, especially with the cyclical nature and demands. What gives you the internal fortitude to keep going?

I say over and over again, I'm not here by accident. I have been around cars my whole life. My dad raced cars, so I feel it's in my blood. I've always wanted to work in the auto industry, and my excitement for this industry has continued to grow over time. You would think after many years that excitement could start to wane, but it's actually the exact opposite. This industry couldn't be any more exciting or dynamic than it is right now. I can't imagine another industry to be working in, and it's my passion that keeps me going day in and day out. I believe if you find your passion, you'll always be excited and engaged to keep going.

Personal Satisfaction

What would you say gives you the most satisfaction in your career?

I've actually spent time away from General Motors more than once and I've always come back and I have come back for the same reason. There are lots of opportunities out there, but the reality is the scale and scope of what GM is working on in this industry is changing the world. There is nothing more satisfying for me than seeing how our products are improving lives in such a meaningful way. At GM we are helping to improve society and our environment, and to me, those are the things that ultimately matter most.

CHAPTER 13

Elena Ford

Chief Customer Experience Officer
Ford Motor Co.

Elena A. Ford is chief customer experience officer. In this role, she leads the organization responsible for creating a world-class customer experience throughout the entire ownership cycle.

In this role, she will work with Ford customer service division and the quality organization to more tightly connect the design and implementation of a world-class customer experience system that connects the interactions between Ford and its customers around the world. In her role, she will continue to be closely integrated with the sales and marketing organization and global dealer network.

Previously, Elena was vice president, customer experience and global dealer. In that role, she was responsible for developing global standards and sharing best practices for planning, training, vehicle delivery, and customer interaction with dealers and within the company. She also was responsible for the development and launch of FordPass and The Lincoln Way, Ford Motor Co.'s ownership experience apps.

Prior to that, she was director, global marketing, sales, and service from February 2009 through February 2013. She was appointed executive vice president, global brand and marketing, Ford Credit in August 2007, leading marketing, product management, and sales support activities for Ford's financial services arm around the world.

Since joining the company in 1995, Ms. Ford served in a number of marketing, brand strategy, and business management roles.

Ms. Ford currently serves on the board of the Edsel & Eleanor Ford House. She previously held seats on the boards of the Henry Ford Health System and The Children's Center.

Ms. Ford is the great-great-granddaughter of Henry Ford and the granddaughter of Henry Ford II. She was born in New York in 1966, where she earned a bachelor's degree in business from New York University.

Education and Lifelong Learning

How do you capture learning opportunities to increase your portfolio of skills? Is this outside your core areas of competency?

Learning is a continuous journey and something that every professional needs to do throughout their entire career. I have spent a lot of time in my career asking questions, seeking answers, spending time in areas that I am unfamiliar with, and finding peers and colleagues who are willing to teach. At Ford, we have created a culture that encourages us to be curious by nature and to learn, discover, and ask questions. Equally important, that curiosity should not be limited to our company.

At the right time it's important to gain higher education learning opportunities such as executive education at Harvard Business School—where you can meet people from other global companies, with different competencies—working together to solve real-world issues. One of my team members was the founder of Sugarpova which proved very interesting in building relationships.

We spend time benchmarking others, looking at some of the best companies in and out of the industry and understanding how we can use their best practices to help drive success in business in key categories.

Work-Life Integration

How have you dealt with work-life integration in your own career?

For most professionals, the management of work-life integration is a challenge, and there is no secret recipe for success. Everyone needs to identify a way to balance that works for them personally. In the beginning of my career, I was not great at ensuring balance between work and my personal life, but throughout my career and journey, I have spent the time reflecting on how to improve the balance and to make sure I was spending more time with my family and children. My focus on improving work-life balance has also had a direct benefit for my teams. Given my appreciation for work-life integration, I encourage my teams to make sure that they, too, are balancing the time between office and home.

Mentor and Sponsor Relationships

How important have mentors and/or sponsors been in your own career? Have they been men or women?

While every career is different, I believe that it is very important to have a mentor throughout a career, and mentors have been critical for me. Most of my mentors have been men and they have been pivotal in exposing me to areas of the business that I wasn't familiar with and helping me learn new skills. I also believe that you only get as much out of a mentorship as you put into it as it is a two-way relationship, and it takes work and dedication. One of my mentors played a key role in pushing me to continue to improve my work and always drove me and the team to be better. While it took a lot of work on my end to prepare for conversations, in return I was able to gain valuable advice.

Taking Charge of Your Career Path

When is it wise to listen to the fear you have of a new job or promotion and when should you ignore it?

Making the decision to take a new job is always an opportunity and sometimes that decision can be wrought with anxiety, especially if it is out of your comfort area. From my perspective, while you need to balance the risks with rewards, my advice is to *always* take the new position even if it's in an unfamiliar area. With a new role in uncharted territory, you have the opportunity to gain new learnings and broaden a skill set that may be beneficial in the future. I always believe that every role becomes a stepping stone for the next job in the company—to learn, grow, and utilize that skill at some later date that will be key for your career.

Resilience

The automotive industry requires resilience especially with the cyclical nature and demands—what gives you the internal fortitude to keep going?

It's no secret that the automotive industry is cyclical, and while the ride down is never fun, the ride up can be exhilarating. And while tough years and environments can be a challenge, resilience is key. I enjoy the work and relationships and strive to make each year better than the rest, and this is a motivating factor in everything I do. It is equally important for leaders to demonstrate their resilience in order to keep teams motivated and focused to deliver the work. If the team sees their leaders pushing to be better, they will do the same. I believe it's important to always stay connected to new information, what's next or new, what will make us better, stronger, more desirable for customers. Ultimately, the success of the company is due to the collective success of the team.

Personal Satisfaction

What gives you joy in your job? What causes you the most angst?

There are quite a few things that drive me and give me joy in the job. The first one is the great teams that I work with today and the teams I have worked with in the past. The collective energy from all the teams around the world is really important and gets me excited. Secondly, there is nothing that gives me greater joy than when we have done the right thing for our customers—whether it is building a great product, taking care of an issue, or going above and beyond—and we hear them say they "love their Ford" or they "love the company." In terms of angst, I worry most when I see people worrying about items out of their control rather than making decisions and moving forward.

14

Julie Fream

President & CEO
Original Equipment Suppliers Association

Julie is the president and CEO of the Original Equipment Suppliers Association. The association represents the voice of automotive suppliers and champions their business interests. Named to the role in 2013, Julie has a 35-year career in automotive, including roles for both original equipment (OE) manufacturers and suppliers in manufacturing, engineering, program management, sales, marketing, and communications.

She is a member of the North American Fiat Chrysler Automotive Supplier Council and the Ford Supplier Council, representing the interests of the automotive supplier community.

In 2016, Fream was recognized by *Crain's Detroit Business* as one of "Michigan's Most Influential Women." In 2015, she was listed as one of the "100 Leading Women in the Automotive Industry" by the *Automotive News*. She serves on the boards of Beaumont Health and Michigan Technological University.

Fream holds a bachelor of science in chemical engineering from Michigan Technological University and an MBA from the Harvard Business School.

Education and Lifelong Learning

How did you learn emotional intelligence and how has this helped you be successful?

After 35 years in the global automotive industry, I have learned that emotional intelligence (EI) is the gateway to successful working relationships. For me, the fundamentals of EI start with good listening skills, specifically—listening to understand. By understanding others' needs and objectives, together you can work toward success. The basis for developing this came by working with customers and striving to balance their company's needs with my own, as well as by working as a direct supervisor. It takes a conscious decision to earnestly listen.

A great lesson in emotional intelligence and one of my favorite quotes by Maya Angelou is "we are more alike, my friends, than we are unalike." Everyone appreciates when we are authentic in our desire to truly understand, accept, and collaborate.

Work-Life Integration

Is work-life integration more possible at some career levels than others?

Rather than look at this as a career-level discussion, work-life integration is typically more associated with the type of work one does. For example, an office worker is likely to have more flexibility in scheduling work and personal time than an assembly-line worker. While that may be true in some instances, I believe we should all strive for work-life *balance*. Each of us should set our own limits to ensure we have time for reflection and personal health. Learning to set these limits is important for both long-term career and personal success.

Mentor and Sponsor Relationships

In your experience, what makes a mentor/mentee relationship fail?

Throughout my career, I have been fortunate to have strong mentor/mentee relationships that continue to evolve to this day. While they vary in formality, length, and depth, they all have these common attributes:

- Defined roles and expectations
- A commitment to open and honest dialogue
- Trust
- Approachability and availability
- Compassion and authenticity

When mentor/mentee relationships fail, I have found that it is due to the lack of one or more of the necessary elements.

Taking Charge of Your Career Path

When do you know that you've reached a pinnacle in a job and need to move on?

In addition to the goals set by an employer, we are also driven by our unique personal objectives in a particular position. These objectives can differ dramatically from person to person and at different points in our career. It is important that we clearly define our objectives, and to the extent that is appropriate, we should also share them with our manager. This is critical for personal development and mutual understanding.

People reach the pinnacle on their job when they attain their desired outcomes both personally and professionally.

When our personal objectives become unattainable or they no longer align with our job, it is likely time to move to a position where there is better alignment. Moving on can mean getting a different role at the same organization or finding a new opportunity elsewhere. With that in mind, I encourage my team to find a worthwhile purpose at work beyond our corporate goals: one that bodes with our life's goals and values. Regardless of what you do, your job should make you feel like you are succeeding.

Most of all, when it comes to knowing when it is time to move on, your intuition can be your greatest ally.

Resilience

What are the biggest drains on resilience in the workplace?

As the saying goes, most people quit their boss, not their job. The biggest drain on resilience in the workplace is when it is not demonstrated at the top. Resilient leaders are trusted, flexible, optimistic, and serve as strong resources to help their employees deal with the pressures of work. The lack of resiliency in leadership drains resilience from the entire organization and is a major driver in employee dissatisfaction and turnover.

Personal Satisfaction

What would you say gives you the most satisfaction in your career?

There are two things that give me the most satisfaction in my career and in my current role. The first is having a positive impact on the professional and personal growth of members of my team and mentees.

The second is working to ensure the automotive industry is a better place for the next generation of employees. While much has changed in this industry over the years, it continues to offer an exciting and challenging career environment.

Sharing what I have learned to help others grow and make a more positive impact on the industry gives me great satisfaction and an even greater purpose.

15

Kara S. Grasso

Vice President, Strategic Operations
DENSO International America Inc.

Kara Grasso is vice president of Strategic Operations at DENSO's North American headquarters in Southfield, Michigan. In this role, she oversees Strategic Planning; Business Operations Planning; Sales Strategy and Systems; Aftermarket; Delivery; and New Entrant Customers. Grasso previously led FCA Sales strategy and activities in all key product areas and DENSO's Product Sales Strategy.

Grasso joined DENSO in 2000 as a senior sales specialist for Chrysler Sales, responsible for leading all Chrysler Engine Cooling Module (ECM) sales responsibilities. From 2004 to 2014, Grasso held various positions within Chrysler Sales, moving up to senior manager in 2011. In 2014, she was promoted to director of FCA Sales with a focus on thermal sales specifically.

In 2016, Grasso was promoted to director of FCA Sales, focusing on sales of thermal, powertrain, engine electrical, body electronics and service products. Her responsibilities include strategizing and strengthening key FCA relationships within purchasing, engineering, cost planning and quality departments, and engaging the FCA Sales associates to create alignment between divisional and product group targets.

In early 2017, Grasso was promoted to vice president of FCA Sales. Her increased responsibility included a larger customer base as the lead of Product Sales Strategy for all product areas. In January 2019, she took on her current role that further expands her strategic oversight.

Prior to working at DENSO, Grasso worked at Freudenberg-NOK as a sales engineer in 1998, responsible for managing all market research data for multi-product sales staff and assisting in the management of the Toyota Sales account. From 2000 to 2005, Grasso also worked for Dale Carnegie in as an instructor for The Dale Carnegie Course, where she taught Dale Carnegie techniques for overcoming fear of public speaking, influencing people and becoming more effective in both personal and professional lives.

Grasso completed both her Bachelor of Science in Business: Organizational Behavior and her Bachelor of Science in Business: Human Resource Management at Miami University in Oxford, Ohio in 1998.

Education and Lifelong Learning

How important is it for companies to create lifelong learning opportunities for their employees?

Companies that create lifelong learning opportunities allow associates to continually develop themselves both personally and professionally. I am fortunate to be at such a company as DENSO, which continually offers training opportunities from entry through executive levels. Most important are the opportunities that allow associates from different regions around the globe to connect within these training courses. Through the actual workshops, dinners afterwards, travel time together, etc., associates build invaluable relationships that easily translate into future partnerships. It is because of these internal relationships that cohesive teamwork can exist.

At DENSO, I have greatly benefitted from these series of leadership development courses built one upon the other. During the capstone of the Regional Leadership Development Program (RLDP), I was pushed to realize and demonstrate the power of vulnerability. This was a really uncomfortable process as I purposefully had kept myself from opening up to my leadership, peers, and associates. I am not quite sure why this was my natural tendency, but looking back I think I was working really hard not to be perceived as the emotional woman in the room. I wanted to have a tough exterior within a male-dominated industry and an even great male-dominated company.

My impression of "vulnerability" was a woman crying in the restroom after an intense meeting or because of an overwhelming workload or a personal issue at home. I thought it was a soft skill that I didn't need to develop. Humbly, I was wrong. I had not connected the magnitude of being truly vulnerable to my trustworthiness. Through this training process, I began to let my guard down. And over time I was able to build trust amongst my working team, my leadership, and my customers. In turn, I was able to realize a positive impact within my team's performance.

Without having these experiences internal to my workplace, I feel the impact wouldn't be the same. The company has invested in me, and, through this benefit, they will get a high return.

Work-Life Integration

How have you dealt with work-life integration in your own career?

I work to feel fulfilled and challenged. I work to help provide a comfortable life for my family. I also work to set a good example of hard work and perseverance for my three daughters Estella, Lucia, and Norah. However, my career and extra projects would not be possible without a true partner.

My husband—who owns his own business—and I ensure our family always comes first. It is not easy to keep this mindset when in the trenches of a major quotation, an internal project deadline, an urgent customer request, or a conference overseas. But what I have learned over time is that I am the only person responsible for keeping myself in balance. I can't allow myself to feel guilty for being at an early afternoon swim meet on a Wednesday when the work day in still in full swing. It is far more important that my girls know I am there to support them.

I also have multiple passion projects that are really important to me personally. These projects, some inside of DENSO and some outside, are important to balance for my own fulfillment. I connect these passion projects to a direct correlation of my mindset in my daily job. Although it creates more commitments on my plate, I am continually developing myself and increasing my network.

Today's technology allows us to be connected 24/7 regardless of where we physically happen to be. The balance challenge now is how to be completely present in your personal life when work is always "on." This is a daunting task for those of us that feel elated by clearing out our email inbox. It is a metal mind game.

Over the course of my career, I have had to push myself to be wife and mom far ahead of employee. When achieving this, I am actually a better employee which translates into a better leader. My team members know that I want the same for them, but that it is up to them to own their personal balance.

Mentor and Sponsor Relationships

Is it important to have a structured system in a company for mentoring or should it happen synergistically or both?

In my experience, mentoring has been more successful when it has happened organically. By more successful, I mean purposeful and sincere. There are many examples throughout my career where an associate has asked for guidance or feedback and through these regular conversations, a natural mentorship has developed.

These types of relationships don't feel forced through systematic bureaucracy in order to achieve a corporate standard. Instead, associates take it upon themselves to seek out mentors in which they feel a connection. Someone's opinion they value and with whom they feel empowered.

My current mentee relationships don't feel like time commitments because I value these relationships as much as the associate. Whether through a lunch meeting or grabbing a coffee off campus midafternoon, I am also able to reenergize myself through these unlikely coaching discussions.

I have also recently recognized that mentor relationships come in many different shapes and sizes. Without a formal approach, your mentorship opportunities are endless. I push myself to learn from everyone I am inspired by—an influential speaker, a peer, a retiring executive, or a young college recruit. If I am inspired, I seek to understand them better and to learn from their experiences.

Last year, a woman sat next to me at a Senior Circle Event. She and I had a great time as we connected first during a Myers-Briggs discussion, then on a book that we were both reading (Influencer). We exchanged business cards and planned a lunch for later that month. Since that time, she has mentored me through a major professional responsibility change, has coached two different workshops for two of my departments, and has had guided me through multiple leadership issues. This unexpected relationship has fostered more mentorship than any company-connected relationship could have.

Taking Charge of Your Career Path

What was the biggest surprise in your career when you faced a new opportunity thinking it was a terrible move, but tackling it in any case—what did you learn and how did it change your view?

After 10 years in the auto industry and pregnant with my third daughter, I convinced myself that I could not have a challenging career and be a good mom. I was overwhelmed with life as it was, and I couldn't imagine increased daily strains of early mornings, long commutes, management responsibilities, and evening family obligations. I also did not have any role models in my office that were making this same scenario work.

During this time, I considered many different career paths that were seemingly more mother-friendly. This was a very difficult decision because I had built a strong reputation and ideally wanted to work within the auto industry. I also had grown to appreciate the family-like connections of DENSO as my teammates were also some of my closest friends. I was at a crossroads, but was determined to make a move within that year.

Four months after my daughter was born, we learned that she had a condition requiring a major spinal fusion surgery followed by six months in a body brace. My husband and I were told that without this surgery, she would not walk by the age of 5. It was an overwhelming time of unknowns.

I will never forget the support I receive from my coworkers, my management team, and my customers. The company supported all my family needs through a temporary part-time schedule presurgery and months off during her recovery. My manager, with a matching blood type, offered to be her primary blood donor. My team came to visit during my leave with food and gifts. It was this sense of community that shifted my opinion of the big-bureaucratic-company versus working-for-yourself career path. I was humbled by all this support.

This circumstance, although awful at the time, taught me an invaluable life lesson. I learned so much about the positive influence of leadership and the importance of the integrity of the company in which you work. I use this example as a sense of empowerment to the young mothers in my organization. When tough times seem insurmountable, you have to recognize the community ready to support you.

*Norah is a healthy and energetic 10-year-old who dances, swims, plays basketball, and is a smarty pants.

Resilience

What are the biggest drains on resilience in the workplace?

Recently, there seem to be drains on resilience in the industry. Maybe it's attributed to the unknowns of future mobility. Maybe it's the ever-looming fear of the next recession. It could be the fact that most automotive companies are operating as lean as possible in order to endure high investments with hopeful big returns while pushing their associates to power shift responsibilities.

What I have noticed is that the grass is not greener anywhere you look. Most of us are experiencing the same jam-packed calendars with little time to celebrate even the biggest wins. When one issue is closed, another is opened. It takes stamina to keep a positive mindset in this industry. It also takes a different kind of leadership.

Generationally focused or not, I have found that we are not only entering a new era of technical development, we are entering a new era of leadership style. Associates want to be heard. They want to be included at the decision-making table. They don't always trust that leadership has their best interest in mind because they see more on paper than they do in action.

I am focused to change this approach at DENSO. Backed by the North America (NA) CEO and along with a team of equally passionate leaders, we are working to create an inclusive culture. Where diverse thoughts or approaches are rewarded, not ignored. We have correlated this inclusive behavior with the amount of innovation we are able to develop over a period of time. This is a significant shift for a company so deeply rooted in Japanese culture, but operating in diverse regions around the globe.

Without the focus on how to properly lead during these difficult times, I believe the resilience of the associate will continue to deteriorate creating even greater mediocrity in output. A company cannot innovate and, in turn, compete in this way.

Personal Satisfaction

What would others say about you or what would you like others to say about you and your influence?

My leadership style is transparent and open. I have found that by sharing company information that is taking place globally, regionally, and locally, the team is more empowered to continue their focused work. They have to be able to connect their role to the bigger mission of the company's efforts. If they cannot see themselves in the future of the company, how empowered can they be to exceed the expectations of their assignment?

My associates would also say that I am their biggest supporter. Over the past couple of years, I have challenged myself to influence career paths to the best of my ability. For those associates that are ambitious and passionate, I will spend countless hours working with them on their *individual development plan*, fighting for them in promotional negotiations, and aligning them with development opportunities.

In the end, I am most passionate about positively shifting the culture of my company to reflect today's new innovation era. Many of the historical characteristics of a Japanese company are beyond impressive: We vehemently support the communities in which we reside. We maintain lifelong employment for strong contributors. We consistently focus efforts towards a safer world for all to live. These admirable traits are much of what has attracted me to DENSO.

Now, as a top leader, I must influence company direction in order to appeal to a more diverse pipeline of ambitious young people. This pipeline is critical to the company's success. I look forward to the day that young women at DENSO have many female leaders who inspire them. This would give me the greatest personal satisfaction.

16

Denise Gray

President
LG Chem Michigan Inc. Tech Center

Denise Gray is president of LG Chem Michigan Inc. Tech Center (LGCMI TC), the North American subsidiary of lithium-ion battery maker, LG Chem (LGC), Korea. In this position, she has overall responsibility for strategic direction, engineering, and business development. She is a member of LGCMI Board of Directors. Additionally, she serves on the board of directors of the Original Equipment Suppliers Association (OESA), a nonprofit trade association that represents the business interests of original equipment (OE) automotive suppliers doing business in North America.

Prior to joining LGC, Gray served as vice president of electrification - powertrain engineering at AVL List, GmbH, in Austria, where she was responsible for leveraging AVL's global capability to provide electrification engineering services to the automotive industry.

Prior to that, Gray was vice president of business development for an electrified powertrain battery startup company in California targeting the China's new energy vehicle market. The majority of her over 30-plus year professional career was spent at General Motors, where she spearheaded efforts in vehicle electrical, powertrain controls and software, including battery systems.

Gray is active in several charitable organizations, including the March of Dimes, where she served as the organization's chair in 2016, 2017, and 2018 for the North American International Auto Show (NAIAS) Charity Preview.

Gray has been a proponent of the academic disciplines of science, technology, engineering, and mathematics (STEM), and is a frequent participant at STEM events. Her strong support of the STEM curriculum played a role in her receiving the 2017 Women of Color Technologist of the Year Award, which recognizes the exceptional achievements of distinguished multicultural women who excel in STEM.

Gray holds a master of science in engineering management of technology from Rensselaer Polytechnic Institute and a bachelor of science in electrical engineering from Kettering University.

Education and Lifelong Learning

How can someone best improve their strategic thinking skills?

Education and Experience! Strategic thinking skills can be enhanced by being exposed to documented examples of industry situations illustrating how a strategic plan is developed and how the results are evaluated. While pursuing my graduate degree, I took a Strategic Thinking class comprised of a number of *Harvard Business Review* case studies. For me the strategic light bulb went *on* as we explored various case studies. Ultimately, having assignments or experiences with an active role in creating and executing strategic product introduction is the best method to enhance strategic skills. So, if you are ever offered that opportunity—*Take it*!

Work-Life Integration

How have you dealt with work-life integration in your own career?

Work-life integration always has been a work in progress for me. A work in progress throughout a 31-year marriage, two adult sons, extended family, community involvement, and a 33-year career. It is only possible due to a determination to have it all and a supportive husband, available family members, and a gratifying career. It requires balancing tasks and sacrificing personal time. Technology has been a great enabler, allowing the ability to perform tasks and communicate from wherever you are.

Mentor and Sponsor Relationships

How important have mentors and/or sponsors been in your own career? Have they been men or women?

I have had great supporters rather than mentors/sponsors in my career. Supporters have been gained or earned throughout my numerous assignments. My supporters are the people who have witnessed by character, intelligence, dedication and integrity as I developed products and supported the growth of my teams. Those supporters have been men and women who have been both in positions of great promotional influence and those who weren't. Supporters of all levels of influence have supported my career growth, which has been driven by both professional promotions and character broadening.

Taking Charge of Your Career Path

What is the best strategy for asking your company for a new assignment?

Asking for a new assignment requires a 'request' strategy and patience. The request strategy should consider the following:

- Status of your project. The best time to request a new assignment is when your current project is in a good or stable state.

- Prepare an itemized list of experiences and skills that you would take to the next assignment.

- Determine what knowledge or experiences you would like to gain from the next assignment.

- In a collaborative manner patiently discuss your next assignment with your manager. Early discussions regarding a new assignment allow your manager time to seek or create new assignment options. Finding the assignment that is a match for you can take time. Mostly importantly, be open-minded to assignments within or outside of your present business unit. Be sure to communicate to management if you are open to both.

Resilience

Is personal resilience built mostly from a person's own internal resources or outside support?

Both! Perhaps 80/20. I believe personal resilience is primarily built on your internal resources. Your character, faith, drive, and determination are internal resources that shape your resiliency. There are outside resources and training that can teach you to recognize your shortcomings and support you in creating strategies to overcome them. Lastly, your support system—family, friends, and professional mentors—provide opportunities and motivation to count all experiences as growth opportunities.

Personal Satisfaction

What gives you joy in your job?

Joy in my job comes from people and product. It is most fulfilling when a group of people come together as team. A team that is greater than the individuals. A team that respects one another and contributes to the best of their ability to bring the best possible product to production. There is nothing better than seeing vehicles on the road you know your team worked extremely hard to execute. People and product bring joy in my job.

What causes you the most angst?

People and organizations that don't respect others or lack integrity give me the most angst.

17

Elizabeth Griffith

Director of Engineering and Program Quality-GM Interior Systems
Faurecia North America

S panning over 45 years in the automotive industry, Griffith's career in the industry has been diverse. In her role as the director of engineering and program quality of Faurecia Interior Systems, Griffith supports a team of 14 people on the GM global account. Some of her responsibilities include growing the business and developing customer relationships; achieving GM and Faurecia safety, quality, and cost objectives; and understanding the complex customer product systems to ensure that the product, processes, and people are focused on success over the product execution timeline.

Before coming out of retirement to join Faurecia in 2010, Griffith held various positions—among them vice president of advanced engineering and program management—at Magna's Intier Automotive from 2001 to 2007. Prior to that, Griffith was the vice president of engineering and program management at Peregrine In.

Griffith's career began at General Motors, where she was part of the first large class of women to go through the General Motors Institute program. After graduating from the program in 1977, she held various positions within General Motors, including manufacturing general supervisor, platform program manager, and chief manufacturing engineer for mechanical components and interiors.

Beyond her professional accolades, Griffith was named one of the "100 Leading Women in Automotive" by *Automotive News* in both 2005 and 2015. She also champions various women in automotive groups, inspiring women to join science, technology, engineering, and mathematics (STEM) fields and careers. She is also the chair of AutomotiveNEXT, inspiring and supporting the next generation of automotive leaders. And was awarded in 2017, the Kettering University Alumni for Management Achievement and has joined the Kettering University Engineering Board.

Griffith holds a bachelor's of science degree in mechanical engineering from General Motors Institute (Kettering University) in Flint, Mich.

Education and Lifelong Learning

How did you learn emotional intelligence and how has this helped you be successful?

After a career spanning 47 years, I've frequently been asked how I survived and apparently thrived in automotive manufacturing and engineering. This question is generally asked in an utterly shocked tone as my career started at GM in 1972, a decade not known for a plethora of women in those fields.

To me, emotional intelligence (EI) has been the foundation of that successful technical career. My EI blossomed concurrently with my leadership style, and it was refined over decades by countless people and problem interactions—from the tough inner city UAW manufacturing floor to the various functional and executive levels and government representatives and multiple workforce generations—a hierarchy of complexity. What worked? What didn't? Why can't I get everybody focused? Obviously, having one older and two younger brothers allowed me to practice EI from an early age. With a dominant older brother and sensitive younger brothers, mama expected me to get things done while she was at work without fights or crying—neither me nor them. In the professional world this is even more important.

The ability to know when and how to pull, push, or just leave someone or something alone; to just say a few words or go in depth with background; or to just smile or catch a glance became a powerful, intrinsic skill. It has been fraught with immense successes (highly effective teams performing virtual miracles) and abject failures (dysfunctional team not even willing to speak to one another or me). The patience needed to develop EI and use it—especially when all I wanted to do was just tell someone to do it, and do it now—has made me not only a much more empathetic leader but more humane.

As an engineer, data is crucial. How do you know you've successfully acquired EI? Well, when people want to work with you and for you, bounce ideas off of you, keep in touch with you, and the ultimate is when they want to follow you, that is when you know you've achieved EI success. They know the authentic you, with integrity and the ability to see them, to have the tough conversations, and to help them reach their potential. You evoke positive potential.

Work-Life Integration

Manufacturing continues to cause a divide in gender—why is this? What can be done to support more diversity?

My General Motors Institute (now Kettering University) co-op sponsor was the GM Fisher Body Fort Street Plant. It was formerly part of the Ternstedt Hardware and Trim

Division and known as the "mother plant" since union pioneer Walter Reuther's mom had supposedly worked there. Walking in those doors in 1972, I knew that I had made a horrible mistake (this started the screaming in the back of my head that continues to this day in highly stressful situations). I should be signing up for my music scholarship classes at the University of Michigan and focusing on achieving my lifelong dream of majoring in music with a minor in library science, not walking into a dark and dirty cacophonous factory with aisle after aisle of men at incredibly large presses (1000T blanking) staring and catcalling.

So from that thought to now? Why do I always recommend that everyone should work in manufacturing? Manufacturing has long had a stigma of being dark, dirty, and not sexy like industrial design, product engineering, or marketing. I've heard so many people express "Why spend all that time getting a degree and waste it in operations/ manufacturing?" Why? Because you learn what's important about the company and yourself. It is a crucible for developing unparalleled leadership skills—fast paced, quick thinking, constant change, and a melting pot of people. The thirteen years I spent at the Fort Street Plant and the subsequent two years later in my career I spent as a plant manager was a transformative experience. The confidence of knowing I might not know something now but could learn quickly, the ability to listen and communicate to a myriad of people, to know I could excel in one of the toughest automotive environments and garner respect while continuing to be true to myself really defined my road to success. Yes that screaming in the back of my head continued (and it does to this day) but experience is a great antidote.

The gender divide can certainly narrow with a three-prong approach. First, as parents, as family, and within the education system, young girls must be taught to compete. They must compete with themselves and boys. They must know the power and confidence boost that comes with winning individually and with a team—to know that stepping into the limelight is okay. They must know that being the best does not make them less. An excerpt from the book *Confidence Code for Girls* as published in *Time*, April 2018, had many salient points. Of note, "that between the ages of 8–14, girls' confidence levels drop by 30 percent." And what I found particularly poignant is that the authors found once the gap is open, most women do not fully recover. I have seen this many times; incredibly talented women undervalue themselves and take a lesser role because they do not think they are good enough so they do not try. They have listened to that voice screaming in the back of their head.

Second, we need to remove the social and economic stereotypes that preclude or sway young women from developing to their full potential. That tough job getting your hands dirty might be the best thing ever. Manufacturing today has some of the best leading edge technology. STEM is not a dirty word. You can be whatever you want and it is okay to make more money than your significant other—you both win. Smart confident women live life to the fullest. And they have taken the lessons learned from the book above, they have learned to "risk more, think less, be yourself."

Third and last, every woman needs to be a vocal, visible role model for those women coming up through the ranks. They should be quick to lend a hand or encouraging word to network. There is a psychological imprint with women that "unless they see it, they can't achieve it." When there is a dearth of women as role models (such as in manufacturing and other technical fields), women find it difficult to believe or even aspire to obtain a position in this field. Fear of the unknown is real. Mama raised all of us with the belief that we could be anything if we studied hard. And with three brothers, I didn't recognize until well into my automotive career that what I was doing was not normal for a woman. During my career of being in so many instances, the first woman, the only woman, one of the critical few executive women in my field, to me it was not special, it was just me plowing through with my job and working hard. Now, reflecting on that attitude, I hope I've set an example for others to achieve more than they ever thought possible—to utilize the power of women's friendships.

Mentor and Sponsor Relationships

In your experience, what makes a mentor/mentee relationship fail?

Being a mentor has been and continues to be one of the most important and enjoyable parts of my life. Like fingerprints, every mentee is unique. I currently actively mentor 18 people. I have been a mentor in both company structured and unstructured programs, and have also mentored outside of my company. To be a successful mentor there has to be a sincere effort on your part to actively listen and to explore a myriad of topics with your mentee. I've had mentees who still seek advice even after thirty years, or others who transition from colleagues to, in a few instances, great friends. It is similar to what I imagine a psychiatrist might do—listen and explore without telling someone exactly what to do, lead them to the realization of what they need to do—be a sounding board. Mentoring works best when there is an easy rapport established; mentoring cannot be predicated on a title within the company—not all vice presidents are successful mentors.

Mentor/mentee relationships fail when the expectation is not clear at the start. A mentor is not going to solve the mentee's functional problem or actively advocate for the mentee. Mentorship works best when it remains above the daily business and focuses on middle- to long-term development and serves as a resource to explore and encourage potentials. One of the worse things you can do as a mentor is utilize your experience(s) or clout to solve a problem for your mentee—they learn nothing other than you are good at, solving problems.

Many of my most successful mentor/mentee relationships have been encouraging young engineers to get out of their comfort zones and into functional areas they deemed too risky (finance, program management, and manufacturing). My failures have been when the mentee was assigned to me, and they wanted me to resolve a problem with their immediate supervisor who was my peer. After multiple sessions it became apparent that their need was not a mentorship. I encourage everyone to be a mentor and to seek a mentor—utilize all resources to improve you.

Taking Charge of Your Career Path

What was the biggest surprise in your career when you faced a new opportunity thinking it was a terrible move, but tackling it in any case—what did you learn and how did it change your view?

The biggest surprise in my career came as the first and really biggest career failure. I had been put in place to clean up a mess: a launch of a major complete interior program in the middle of building more than 200 large injection tools and installing millions in equipment. The internal team was in disarray, and the customer was not happy. There were daily escalation calls at the executive vehicle chief level. This was a situation I had dealt with many times before and it was a forte. Initially, great progress was made. But during the final weeks of the launch, I let dimensional issues become adversarial with the customer; it became very personal. I was removed, and I was devastated. Yes, tears happened, and I was determined to quit. Obviously, no one really appreciated my hard work.

It took me a couple of days to return to rational thinking (and a quick discussion with my mentor). I've never been a quitter—and a tenet of my leadership inner voice (when it is not screaming) says "don't let them see you sweat." My team also needed to see I was okay—they were about to mutiny. So I went from a team of 18 to a team of me—as a staff assistant to the engineer-in-charge. No direct reports, no direct authority. It turned out to be one of the best jobs I ever had. The access to upper leadership, developing presentation and speech writing skills, getting people to do things with implied authority, organizing and facilitating executive staff meetings, understanding how engineering impacted other functional groups—I learned more than I thought possible. And the unexpected result, after just 18 months, was a promotion to the executive ranks.

The lesson learned was very clear to me: while it is fine to have a healthy ego, exercising it with the customer is not. Hubris does not work.

Resilience

Do you consider yourself resilient? If so, how did you become that way?

Adaptability, tenacity, and an unerring sense of fun aid in my resiliency. Life is too short not to have fun, and I chose to be happy every day.

Resiliency started for me growing up in Southwest Detroit where daddy had moved us from South Carolina for a job at Cadillac. Daddy died when I was young, and mama struggled to raise my three brothers and me—grannie wanted us all to move back South. We stayed and mama got a job but it didn't cover everything. So there was food stamps, Goodfellow Christmases, and early morning paper routes. The prevailing attitude, however, was always "we will do the best we can" and there was always laughter.

My role model for resiliency was mama—work through issues calmly, and if the original plan doesn't work or you can't control it, have confidence that something else will work. And when in life you lose someone who causes you immense joy, know that time does heal the pain and memories are always there. A huge drain on resiliency is never letting go of the past. So many people dwell on looking back with regret that they lose sight of the road in front of them.

I have a couple of coping mechanisms. My teams, for instance, have always appreciated my calm demeanor and urging to "let's just take a breath" (yoga does help). I've also found that taking a stressed-out person through a short visioning of "what is the worst that can happen" tends to put things in perspective. And what has been a foundation for me is that I can have stress at work or stress at home, but not both. Luckily I have an absolutely fantastic husband of 31 years who has been a successfully retired United Automobile Workers (UAW) member for 19 years. He keeps me grounded and laughing every day. I have two special nieces who attempt to keep me young and edgy. And when my resiliency reservoir needs replenishing, there is always a long Harley motorcycle ride (yes on my own bike) with my husband and friends or a quiet morning reading in the sunroom with my Birman cats snuggled close.

Personal Satisfaction

What gives you joy in your job? What causes you the most angst?

When I see people working together to achieve personal and professional goals while having fun—that gives me joy. The burgeoning autonomous and electrified technology has just added an additional layer of interest for me. My automotive journey has been spectacular—the people I've met, the places I've been, and the tremendous technology changes. But the most joy I've derived is when someone I've met along that path delivers a simple thank you—"you took the time to encourage me, and I became an engineer," or "I'm working in a plant and I love my life," or "my dad said you were the best, and I'm going to be an engineer like you," "you have really inspired me." Faurecia has afforded me the tremendous opportunity to spearhead leadership, diversity, and STEM topics in various forums. It's something I enjoy, and it's critical to the sustainable future of automotive/mobility to engage young people today. To share personal stories and get them excited about automotive.

Do I have angst? A prior question answered this. Every day I chose to be happy. There is no room for angst, only a confidence that the future is a road to be travelled and enjoyed…with curves and switchbacks galore.

18

Mary Gustanski

Senior Vice President and Chief Technology Officer
Delphi Technologies

Mary Gustanski is the senior vice president and chief technology officer of Delphi Technologies, formerly Delphi Automotive. In this role, Gustanski is responsible for the company's innovation and global technologies, including advanced propulsion systems for future vehicle electrification. Prior to this role, she served as vice president of engineering and program management for Delphi Automotive, which spun off its propulsion business to become Delphi Technologies in 2017.

Ms. Gustanski began her 37-year career as a college cooperative student at the former AC Spark Plug Division of General Motors. She was hired full time as an associate manufacturing engineer in April 1985 and went on to hold several positions in engineering and manufacturing, including senior manufacturing project engineer, senior project engineer, and production superintendent, prior to being promoted to divisional plant quality manager in 1997. Ms. Gustanski became chief engineer at Generators in 1998 and was promoted in 2001 to technical center director. In 2003, she was named global director of manufacturing engineering.

In 2006, Ms. Gustanski was appointed as a member of the powertrain systems executive staff as the divisional director of engineering, customer satisfaction, and program management. She was named as Delphi powertrain's vice president of engineering, operations, and customer satisfaction in 2012 and, in August 2014, was appointed to the corporate engineering team as vice president of engineering and program management.

Ms. Gustanski was recognized as one of the "100 Leading Women in the North American Auto Industry" by *Automotive News* in 2010 and 2015.

Education and Lifelong Learning

How have you structured your own approach to lifelong learning?

I've approached lifelong learning as a journey, not a destination, believing there is always more to learn and with every task presenting an opportunity. Certainly, formal training cannot be replaced, especially in the more technical areas of our industry, but I prefer to think of this as foundational. My degrees (bachelor of science in mechanical engineering with electrical option and master of science in manufacturing management) provided the basis of my engineering knowledge, structured problem-solving, and management capability to prepare me for my career. Ultimately, however, it was each and every work assignment and personal challenges, such as my first executive assignment in a mostly male-dominated field, that filled my toolbox of knowledge.

All of my international experiences have also been very influential to my lifelong learning. From the interactions during visits to my manufacturing plants and technical centers around the world to the very interesting customer dynamics that I've experienced in this global industry, I've adapted my leadership style to accommodate various cultures. I continue to be touched by each and every encounter.

With each new position I held came more challenges but, more importantly, significantly more opportunity to expand my knowledge. Bottom line, lifelong learning can only be accomplished when you truly tackle every challenge presented—including those outside of your area of expertise—and search for opportunities that enhance your skills, increase your knowledge base, and allow you to ultimately become a better person.

Work-Life Integration

Technology has enabled the move from work-life balance to work-life integration. Do you think that the pendulum will swing back to more separation?

I can't believe that the pendulum will ever swing back to more separation. Yes, technology enabled the move to work-life integration, and from my standpoint that is a great thing! When I think back to the early days of my career, you had to choose between work and personal because, at that time, you had to be physically there. Now, with technology, I can be productive at any time and from anywhere. Technology has been transformational for managing a balance. However, the risk that must be avoided is swinging the pendulum to the point of using technology to work 24/7, causing you not to be present in your personal life because the tools available create a constant distraction.

As we launched Delphi Technologies (a split from Delphi Automotive) a year ago, we embraced an initiative to create a culture that achieves results while making the company more human. To accomplish this, we initiated coaching for our entire leadership team, and one of its main principles was to "Be Here Now." The premise is that in whatever you are doing, conference calls, one-on-one discussions, family dinners, or personal conversations, you need to be focused on that moment—and the people in them—and not multitasking. I now make it a point to close my laptop and put away my phone when I'm interacting with one of our technologists, giving my full attention to the person talking and the topic of discussion. As a result, I've found the meetings to be much more productive; and I'm more engaged than ever with our people, which is very rewarding. I think this is a critical point for all of us: Use technology for its flexibility, but not at the risk of shutting down other essential areas of your life. More importantly, Be Here Now to get the most out of each and every interaction, in both your career and your personal connections.

Mentor and Sponsor Relationships

Is it important to have a structured system in a company for mentoring or should it happen synergistically or both?

I believe a structured mentoring system is a necessity for all companies. First, it is important to define what a mentoring program really is and then to educate your team on how to use it. Second, it is critical that HR systems support the mentor/mentee with a process to guide the discussions and realize the benefits; otherwise, the meetings are happening just to "talk." Finally, if done well, a mentoring program should realize great benefits in developing future leaders to keep a healthy progression and succession program.

That said, once a company has implemented a successful mentoring program, the benefits will grow exponentially. As more employees understand the purpose and value, mentor/mentee relationships will form naturally and happen more informally, providing great benefits to all involved. Additionally, as leaders become more effective mentors, they will be stronger contributors to the company.

Having been on both sides of the program, as a mentee and now a mentor, I see strong benefits for both roles. My mentors along the way were great sounding boards and provided insightful advice to help shape my leadership style. Now as I mentor to many individuals, helping to guide them through both career, as well as personal choices, it makes me more aware of the team challenges. This in turn has guided my leadership interactions with the broader team. The benefits of a mentor/mentee relationship are significant and a structured program is the best method to reap the benefits.

Taking Charge of Your Career Path

What intentional decisions have you made about your career and were there opportunities you received that you had never considered?

One of the most important decisions for my career was made early on as I embraced the advice of a top leader. As a cooperative student, we had weekly meetings that included visits from managers to discuss various business and career topics. When I was a freshman, our general manager of the division was the guest speaker and he made a very strong point that sticks with me yet today: "You can never have enough education and take every opportunity offered, even if it's not on your short list of choices." Those words had a significant effect on me and have stayed with me throughout my career.

A good example of this was when I was making my degree choices at General Motors Institute of Technology (GMI) (now Kettering University). I was intrigued that they offered an electrical engineering option to my mechanical degree. At the time, there was no limit on credit hours for the semester payment, so an electrical engineering option allowed me to learn both aspects of engineering for the same price. What a deal! And I have to admit that general manager's comment was in the back of my mind as I signed up for a degree that was commonly known as "ME-suicide," because it required a significant amount of credit hours and effort to complete. Now, seeing the automotive transformation that we are in the midst of, I know it was the right choice, creating for me a mechanical and electrical education foundation that I draw upon every day.

During my career, there were definitely opportunities I received that I never considered were for me because they weren't traditional from an engineering standpoint. One such example was when I was asked to lead a special assignment designed to reduce the warranty rates of one of our products. While the work played to my strengths, the reporting lines—to both the operations manager and director of engineering—was new to me. It wasn't something I was sure I wanted to tackle. I'm glad I did. I learned how to adapt, figuring out how to fuse the direction provided by each function into one so that their needs were met, *and* I successfully achieved the desired targets.

Every role aligned with my unique education, leadership skills, or next stage in overall growth and development. As every one of these opportunities was presented, I can recall thinking, really? But then very quickly thinking, why not? I love a challenge and I can't pass up this opportunity!

Resilience

The automotive industry requires resilience especially with the cyclical nature and demands—what gives you the internal fortitude to keep going?

Yes, automotive certainly requires resilience given not only its cyclical nature but its volatility. In my career, I've experienced two company splits (first from General Motors as Delphi and then from Delphi Automotive as Delphi Technologies), as well as a significant bankruptcy period. I've watched vehicle brands come and go, as well as whole OEM companies either merge or completely disappear. I've personally stood up operations, and then had to close them. Couple all of that with things such as a rapidly shifting regulatory landscape, the dynamics of globalization and the political and societal factors that accompany it, and the advancement and acceptance by vehicle owners of new technologies—and you can see why our industry is not for the faint of heart.

However, that is what keeps it exciting.

I'm often asked why did you not leave the industry, given your knowledge and capabilities? The answer is simple. Every time I felt myself reaching a career plateau, a new challenge would come along. Yes, some were more difficult than others, but ultimately, I was being given the opportunity to make a difference. And isn't that what it's ultimately all about? To me, this is essential to getting out of bed every morning and doing what we do. To take an exciting challenge, make it yours, and make it count.

Now, as I look back on the past 50 years in our industry, those challenges pale to the transformation we'll see in the future. For me, there couldn't be a more exciting time to be in the auto industry. This next period of accelerating innovation is what drives me, and our team of technologists, as we invent the future of propulsion.

Personal Satisfaction

What would you say gives you the most satisfaction in your career?

That's easy—it's all the careers I've influenced along the way! From students to production employees to young engineers and more mature team members, each and everyone has truly touched me by their willingness to learn, succeed, and contribute to the future of our industry and our company. And most importantly, we all know when you teach, you learn.

My greatest learning over my years of experience has to been to slow down and listen to the feedback. My strong personality and desire to drive results, often is accompanied by impatience with the team. So I've made it a priority to listen intently to my peers and employees prior to reacting. It's difficult; but necessary. It's amazing the amount of information and knowledge you acquire by listening to the feedback, not to mention the respect and trust you generate with others. For me, circling back to an earlier question, this is how I've achieved a lifelong career of learning!

19

Marcy Klevorn

Executive Vice President and President, Mobility
Ford Motor Co.

M arcy Klevorn is executive vice president and president of Mobility, Ford Motor Co., effective June 1, 2017. In this role, she is responsible for overseeing Ford Smart Mobility LLC, which was formed to accelerate the company's plans to design, build, grow, and invest in emerging mobility services, as well as global data, insight, and analytics. She also chairs the board of Ford Autonomous Vehicle LLC formed in July of 2018. Klevorn reports to Jim Hackett, Ford president and CEO.

Previously, Klevorn was group vice president for information technology (IT) and chief information officer, a position to which she was named in January 2015 when she also was elected a Ford Motor Co. officer. In this role, she has overseen the complete transformation of the company's IT tools and talent to put Ford in the forefront of technology companies globally.

Klevorn has spent her entire Ford career in IT, serving in a variety of positions in the Americas, Ford of Europe, and Ford Credit.

She joined Ford in 1983 in telecommunications services and worked at various positions within Ford IT and Ford Credit through 2004. In 2005, she was appointed product lifecycle management global director and implemented process changes in data and information management across product creation.

In 2006, as enterprise defragmentation director, Klevorn led the strategy and implementation of infrastructure defragmentation, data center consolidation, and overall systems management at Ford. From May 2006 through September 2011, she led Ford IT operations.

From September 2011 through September 2013, Klevorn served as IT director for Ford of Europe and was a member of the Ford of Europe operating committee. She then was named director, office of the CIO, responsible for managing Ford's global IT business applications, architecture, data centers, web-hosting requirements, engineering, and infrastructure services.

Klevorn holds seats on the boards of Lawrence Technological University and Pivotal, a cloud-based software technology leader. She was born in 1959 and earned a bachelor's degree in business from the University of Michigan, Stephen M. Ross School of Business, Ann Arbor, Michigan.

Education and Lifelong Learning

How important is it for companies to create lifelong learning opportunities for their employees?

It is immensely important for companies to create lifelong learning opportunities for their employees. In fact, I am struggling to think of anything more important with the exception of a healthy culture—of which learning should be a part. The pace of change has never been more rapid and only shows signs of escalating. While this is true in virtually all industries today, the automotive industry may have one of the fastest rates of change. Our industry is being disrupted by participants like Tesla and competition from nonautomotive tech companies whose business it is to move at lightning speed. Without the benefit of lifelong learners, the probability of a company keeping up with today's pace of change—which means constantly learning new skills and expanding the scope of your gaze as industry lines blur—will be close to zero. One real example is the position I hold today leading Ford Mobility. This organization did not even exist two years ago! Learning everyday, innovating, and being willing to disrupt your own business are essential to survive now more than ever to avoid being left behind with a business that has been commoditized.

Work-Life Integration

Have you ever chosen work over family or vice versa? Has this gone well? What did you learn?

When I first looked at this question, my initial instinct was to answer that I always have chosen my family first. Upon reflection, it is true that I have chosen my family in big ways, but not always in small ways. For example, when a life-changing decision had to be made, I chose my family always. I turned down relocations that were also promotions when it was not the right time for my family. I worked from home a few afternoons a week during a time when doing so was not at all common and was a risky thing to request. I made these choices even if it may have been interpreted as not being committed to my work and company. It is important to point out, however, that the trade for me in these cases has always been accountability—with every decision, I am accountable and no one else. So if, by turning down a promotion, I risked not being asked again, I was willing to make that trade—the decision was on me and no one else. If I asked for flexibility in my work location or schedule, I was accountable to deliver as seamlessly as ever. By doing so, I learned that I would be asked again, that my flexibility would not affect my ability to ensure success. I also learned that we need to be confident in our own abilities, confident that our work ethic and abilities will be recognized, and confident that a company that shares our values will work with us during life's journeys. As I said earlier, however, it's different when we are balancing the smaller decisions. In those moments, we make day-to-day trades to do the best we can for everyone who is counting on us. In some of those moments, I have chosen work over my family. Sometimes this meant being late for something or even on the rare occasion missing an event, but we make these trades instinctively to deliver work we are proud of, to handle an urgent issue, or to give time to a team member who needs time. For many of us, these small trades, while unavoidable, also breed creativity. Over the years, I have become very creative about how I integrate work and home life and I continue to innovate in this area daily!

Mentor and Sponsor Relationships

How important have mentors and/or sponsors been in your own career? Have they been men or women?

Mentors and sponsors have been key in my career, even when I did not realize I had them. The opportunities for formal mentorship, with regular mentoring meetings and the like, did not come along very often, but when they did, I took advantage of them. As I look back, however, these opportunities were not something I sought out myself. In my experience, there are mentors all around us, and that is where we can find some of our biggest learnings. It can be as informal as watching how effective people work and, importantly, how they treat people. We have the opportunity to learn from everyone at every level and, sometimes, learning means seeing what does not work. On sponsors, I was fortunate to have both men and women act as sponsors in my career. Most often, given the demographics of our industry in the past, these sponsors were men. These individuals provided great feedback and opportunities to stretch. They also recognized abilities in me that I did not see myself, and had faith in me to try new things. My sponsors supported me when I was not in the conversation, expressing their confidence in me with others and creating more opportunities. I know I would not be where I am today without the support of these individuals. I am forever grateful and work hard to pay it forward.

Taking Charge of Your Career Path

When do you know that you've reached a pinnacle in a job and need to move on?

I know it is time for me to move on to a new role when I start feeling restless. Usually this comes when the change or transformation that I was driving is complete, although those days that may not ever come around again given the pace of change in our world!

While there are usually new things to learn in any situation if you look hard enough, I also have felt restless as my learning curve in a position has flattened out. This creates an opportunity to be curious about other areas that I may not have had time to reflect on before. It is enjoyable to dig in and tackle new territory, to generate a healthy dose of discomfort—healthy stress—and deal with ambiguity. It also is important to move on when it becomes clear that another pair of eyes could better see things that my experience may cause me to miss. It is great to have someone with complementary skills come into a position after you and see the opportunities to optimize for something different.

Resilience

The automotive industry requires resilience especially with the cyclical nature and demands—what gives you the internal fortitude to keep going?

I work for a company that shares my values. I know that we are in it to win and we will. I believe in our leadership. I believe in our approach as a family company, and I believe in the family of employees that make up our workforce. Being in automotive does require enormous resilience, but being able to believe so deeply in the mission and the people around you is what creates the fortitude to keep going. Also, I am a fourth generation employee of Ford Motor Co., which makes my family and the company I work for inextricably entwined. The knowledge that the most important thing in my life is my family also helps me stay centered during any storm.

Personal Satisfaction

What gives you joy in your job? What causes you the most angst?

The most joy in my job comes from working side by side with our team to develop a strategy and deliver on it, all while creating a strong, healthy, and fun team dynamic. Delivering as a team is the most satisfying victory of all for me, far better than any individual win. We spend more time at work than we do anywhere else, and having been able to learn from others, grow my network of relationships, and enjoy the journey is pure joy. I experience angst when a team member is unable to put aside self-destructive behavior, despite our best efforts to work together. I feel strongly that it is my role to ensure our team members' success. When, for whatever reason that is not possible, it is painful.

Alisyn Malek

COO and Co-founder
May Mobility

Alisyn Malek is the COO and co-founder of May Mobility Inc. She was formerly the head of the innovation pipeline at General Motors. Prior to that role, she was an investment manager at GM Ventures. There she led investment in the autonomous space, including the early negotiations with Cruise Automation, helping to cement GM's autonomous strategy. She has experience as an automotive engineer, having led a global team to develop advanced charging technology for GM's Spark and Bolt EV products. Malek was recognized as a top ten female innovator to watch by *Smithsonian* in 2018 and named a top automotive professional under 35 to watch by LinkedIn in 2015 for her work in cutting-edge product development and corporate venture. Alisyn holds two bachelor's degrees from the University of Michigan, one in Mechanical Engineering and one in German language, as well as a master's degree in Energy Systems Engineering from the University of Michigan and an MBA from the Kelley School of Business at Indiana University.

Education and Lifelong Learning

How do you capture learning opportunities to increase your portfolio of skills? Is this outside your core areas of competency?

In order to increase my portfolio of skills, I first thought about what kind of work I most enjoyed and what skills were needed for that type of work. Working backward from that framework, I could identify which skills I would need to get there. I learned that you can get very creative in how you learn these new skills and that not all skills are developed "on the clock" during your day job.

I knew that I wanted to be a leader, and I knew I wanted to focus on the entrepreneurial sector and building businesses. A major part of starting a business is being able to form a network, so I worked through volunteer activities to improve my relationship-building skills. As an introvert, networking didn't come naturally to me, but the practice was key. It helped me continue to be successful in roles where networking is crucial, like venture capital and raising a company from the ground up.

I also knew I wanted to learn more about creating a brand and getting people excited about it. I used my experience from organizing an artist collective to help me build a brand. It was that experience, combined with my work as an engineer, that helped me land my role in venture capital, putting me one step closer to my goal of working in the entrepreneurial sector.

Work-Life Integration

Is work-life integration more possible at some career levels than others?

Work-life integration comes down to boundaries, some of which you can set yourself and some that you cannot. It may seem like this is easier to do at a more senior role, but there are often multiple factors at play including the nature of the work, the individual, and their surrounding team members. For any leader, especially as they get higher into the organization, interruptions to work-life integration can be inevitable. This is where clear prioritization becomes really important. Good team members understand when to delegate and are clear with leaders and team members on what specific types of work are likely to be delegated.

As an individual, when both your superiors and your direct reports understand your framework for handling work when it starts to pile up, it is much easier to balance work with life. The times where this balance tends to fall on the wrong side for me is when I do not feel like I have a say in where my time is going and the ability to plan accordingly. I struggled with this a lot when I was an engineering business manager working as the go-to person for an executive. Their schedule was unpredictable, which meant accepting mine would be too.

As I looked for roles after that one, I kept in mind that I wanted more control over my schedule and searched accordingly. I landed in GM Ventures, where my investment deals drove a lot more of my schedule. Of course, there were still unpredictable things that would come up, but, on the whole, I got to balance my work schedule and travel more. This imbalance can happen once in a while, but when the frustration becomes systemic I try to reevaluate what is not working so I can fix the situation or find a new role with better balance.

Mentor and Sponsor Relationships

Did you benefit most from forming mentor/sponsor relationships or from other relationships or networks?

I view my whole network as a mentor network. When you realize that there is something you can learn from every person you meet, it opens your eyes up to opportunities for improvement and how much you can learn from your relationships. My mentor network has some people whom I consider more formal mentors, and I try to check in with them on a regular basis, but I also continue to see the other individuals in my network as situational mentors. When I encounter a challenge, I think through my whole network to pinpoint who might be able to help me sort through the situation and provide some experienced insight.

I also rely on sponsors to keep me involved in conversations that I otherwise would not have access to. Keep in mind as you engage with others that you do not always know who your sponsor will be. I was introduced to cofounders of May Mobility by a mutual contact who was not even aware of my job search or interests, they just thought I might be a good fit—and they were right. I hadn't categorized them as a sponsor in my mind, but making that strong impression helped them to champion me without me even knowing it!

Taking Charge of Your Career Path

What intentional decisions have you made about your career and were there opportunities you received that you had never considered?

Most of the moves in my career have been very intentional decisions around trying to gain a deeper understanding of business and grow my skill sets, or to get me into a safe space where I can reevaluate the landscape and choose my next direction.

I left what was seen as an upward trajectory position in engineering because I was focused on finding more entrepreneurial positions and jobs where I would be a better cultural fit. An opportunity came up to join GM Ventures, but I assumed I would never be considered because I did not have a business background. When I interviewed and got the job I was elated. I felt like I had made significantly more progress toward my goal of working in the entrepreneurial and new tech space than I would have if I had gotten the expected promotion in a division where I was not a great fit.

I intentionally leveraged my position in GM Ventures to build my network and search for opportunities to join other tech companies and startups. Soon after, a contact in my network reached out to me about joining May Mobility as a cofounder. I was probably the only person to whom that seemed the obvious conclusion, but I had worked throughout my journey to be clear with myself about the goals I had set out to achieve, and this opportunity would put me there. I left my corporate job, turning down attractive offers from other tech companies, and set off on my current career adventure.

Resilience

Is personal resilience built mostly from a person's own internal resources or outside support?

You absolutely must have your own internal resources for personal resilience in order to manage the ebbs and flow of a day or a week, but it is key to understand what helps build your resilience. It might be things like having a creative outlet to blow off steam or time with a mentor or confidante that helps to re-level your mood and continue moving forward.

I build resiliency through activities like running or creative pursuits. I was an athlete growing up, and I find that pushing myself physically, in a way that I can control, has helped prepare me mentally to handle challenges that require me to be more resilient. I'm also a creative person, and I find that having time to make music or visual art uses a different part of my brain and reduces the threat of burnout.

I also rely on my family and team members to help when I struggle. Additionally, my mentors have been a great resource when I find my resiliency being tested, especially when I find myself in entirely new situations. When trying to push through an engineering project that would keep an entire launch project on time, I was able to reach out to leaders in my company to get guidance on how to make it successful. That experience really helped reduce my stress and rebuild my resilience because I knew I was on the right path with a supportive team. Moreover, I think it is important when I ask for others for help that I try to make sure I am supporting them as well when they need it. We all have challenges in life that test our resilience, and it is important that we are actively guiding one another to better versions of ourselves.

Personal Satisfaction

What would others say about you or what would you like others to say about you and your influence?

Across the board I think people consider me reliable; if I say I am going to do something I will make sure it happens. As I have grown as a leader, it has been important to me that I create a system of transparency. I want to create a culture where everyone is able to make the best decisions for themselves and their work.

Leading with this transparency allows me to more easily trust my team because we can have open and honest dialogue about what is working and what is not. When I was working on advanced charging systems we had a lot of challenges that prevented us from being able to deliver the final solution in time for launch, but I was always clear with my team, understood their challenges, and was able to put together a plan for the vehicle launch team that would meet their needs. I was always clear about what I could deliver and made sure we had alternative plans easily understood to all team members.

With time, this trust increased like a savings account with both peers and leaders throughout the company. Having that bank of trust meant that when I needed to ask for help, people at all levels were very likely to pitch in. With this approach, I think I am able to get the best out of my team and inspire people to do more because they feel like they get to live up to their best potential and trust that I will be their support whenever they should need it.

Carrie Morton

Deputy Director
Mcity

Carrie oversees day-to-day operations of Mcity, the University of Michigan's public-private partnership devoted to advancing the development of connected and automated vehicles. She is actively involved in supporting strategy development and execution, and fosters collaboration among Mcity's industry, government, and academic partners.

Prior to joining Mcity, Carrie served a dual role at the University of Michigan Energy Institute. As director of business development, she helped broaden industrial relationships with energy faculty. Morton also served as assistant director for collaboration and industry outreach for the US-China Clean Energy Research Center-Clean Vehicle Consortium.

Carrie joined the university in 2011 after more than a decade in the automotive industry, primarily with the Robert Bosch Corp. In her last role at Bosch, she was manager of government projects and responsible for leading publicly funded research projects, with a focus on engine combustion.

Carrie holds a bachelor of science degree in mechanical engineering and a master of engineering degree in automotive engineering, both from the University of Michigan.

Carrie enjoys spending time downhill skiing and sailing on Lake Michigan with her husband and two sons.

Education and Lifelong Learning

How important is an MBA or other graduate degree?

Graduate degrees can be very valuable in your career, especially when they further your knowledge in a field of interest to you. When considering an advanced degree, it is important to think about why you want to invest your resources—time, energy, and money—toward that effort. Will you be fulfilled by growing your knowledge in this area? Are you moving toward your passion?

If your main reason to pursue an advanced degree is for a specific career position, it may not fulfill you in the ways you expect and in the end may not be a wise investment. That said, continuing education can introduce you to a community of experts and other learners in your current field, or a new one, which can be inspiring and energizing along your path.

It may seem obvious to pursue advanced degrees in established tracks such as engineering or business. Building your leadership skills is just as important. Seeking opportunities through your employer or outside organizations, such as professional societies, nonprofit boards, etc., can be a great way to find leadership learning opportunities.

The course I have followed in my career has been less about advancing in a straight line and more about satisfying my curiosity for exploring uncharted territory and my motivation to make a difference. My degrees have been important for the growth and knowledge they represent as much as the diplomas themselves.

Work-Life Integration

How have you dealt with work-life integration in your own career?

As a mother working outside the home, work-life balance has always been an important consideration in my career decisions. You must have boundaries. It is not easy. Your family must be part of your support network. And you need to recognize that there are tradeoffs. It's unrealistic to think you can always be fulfilled at work and at home simultaneously. Sometimes, you have to focus where you are needed most.

After my first child was born, I took a big chance on a new career role. I stepped away from hands-on engineering and accepted a position in engineering process management for less travel and better work-life balance. I wanted schedule flexibility so I could be the best mom possible, but I worried about where this move would lead. Was I dead-ending my career? What if this new role was not as fulfilling? In time, the new skills I developed allowed me to move on to managing publicly funded research projects. This career pivot led me to something far more fascinating than I could have imagined.

Years later, when serving as the public face of Mcity, there were many more nonnegotiable demands on my time and much less flexibility in my schedule. I learned that by growing our team and their skills over time, and sharing responsibilities, I was better able to balance my work and my personal life, and so was our team. You have to learn how to effectively lead a team, and trust your team, because you can't handle everything on your own, and you shouldn't.

A healthy integration of work-life is always possible, but only if you make it a priority.

Mentor and Sponsor Relationships

How important have mentors and/or sponsors been in your own career? Have they been men or women?

I am grateful to have been mentored by both men and women and continue to benefit from these relationships today. I am lucky that my mentors found me. It wasn't until I was well into these relationships that I even realized I was being mentored, although not through a formal process.

Early in my career, as a female in a male-dominated field, I had two male mentors, both seasoned engineers and auto industry experts at the pinnacle of their careers. They still found time to guide a young engineer with a passion for cars. Their support, advice, and example have shaped my career in immeasurable ways.

Later in my career, the value of mentoring did not diminish as my needs shifted from technical expertise to empowerment and decision-making skills. In recent years, several women have been my advisors and mentors, helping me to grow as a manager and a leader. Their counsel has been invaluable in learning how to navigate corporate culture and other challenging situations.

The wisdom of others with more experience helped provide me with perspective and a long view for making difficult decisions and choices. Looking back, the one thing I would change would be to pursue mentor relationships in a more proactive way. I continue to value the advice of my mentors, and I try to live by their example by helping others along the way.

Taking Charge of Your Career Path

How do you evaluate when to take career risks?

I think of my career path as an interesting zigzag guided by my internal compass and often influenced by my stage of life and my family. When making decisions, it is important to follow your personal compass—a set of values that will help you navigate the landscape. We all have niggling fears and self-doubt. If you stick to your compass you will sense when an opportunity may not help you achieve one of the points on your compass and is something to investigate and consider further. If you can find reasonable solutions that will keep you on course, then you can proceed with a healthy amount of preparation. If not, the opportunity is probably not right for you.

These are the four values I pursue in charting my path:

- Make a Difference—I must feel I am spending my day making a better future for my kids.

- Foster a Great Team—I work to create an environment of mutual respect where the team all believes in our mission and sees the benefit of their work.

- Balance Work-Life—When the scale tips too far in one direction, adjustments must be made. It has taken a lot of courage to keep that balance even under extreme career demands. I am still practicing.

- Gain Knowledge—I have a passion for advancing my skills and learning new ones along the way.

Resilience

How do you build resilience in a team?

Create a diverse and inclusive work environment. Building a work culture that is inclusive of the whole person, and brings together people with different strengths and talents, provides endless rewards to me personally while also benefiting the organization where I work. It enables people individually to be creative and allows for learning opportunities. In my experience it also leads to solutions where otherwise teams that are homogeneous in their backgrounds and ways of thinking may get stuck. Building a team where relationships are based on respect and people see the value of their work, leads to better results.

Apply emotional intelligence. Nobody wants to work with or for a jerk. Empower people to do their jobs without micromanaging them. Let them know you support them, even when mistakes are made. Be kind when personal situations outside the office, such as family emergencies, temporarily impact performance. People don't leave jobs, they leave bosses. And toxic individuals ruin productivity and therefore the bottom line.

Ensure workloads are balanced. Make sure every member of your team can experience the ebbs and flows of work. If people constantly feel in crisis mode, they are not as willing or able to reach out and help others, hindering the group dynamic. Eventually, even heroes have their limits. A team's strength comes from a healthy balance rather than heroic measures. Resiliency is a mindset, but also is most achievable by a team when reinforced at an organizational level.

Personal Satisfaction

What would you say gives you the most satisfaction in your career?

For me, career satisfaction comes from finding a harmonious balance between my four personal compass points. I don't feel fulfilled unless I'm part of a supportive work team, have a healthy work-life balance, and can pursue new knowledge that leads to using my skills to make a difference.

I also gain great satisfaction from forging new paths. Pioneering new discoveries and solutions inspires me. Earlier in my career, my focus was on finding technical solutions to engineering puzzles. Today, my satisfaction comes from knowing that our work at Mcity, at its core, is about using technology innovations to make transportation safer, greener, and more accessible in ways that will benefit not only my children—one of my compass points—but society as a whole. Publicly sharing what we learn from our work can help others working in intelligent transportation across the world.

Alisa A. Nagle

Head of Human Resources, FCA-North America

A lisa A. Nagle was named head of human resources, FCA-North America, in April 2019.

Previously, since 2015, Nagle was chief human resources officer for Stoneridge Inc. In this position, she was responsible for leading a global workforce focusing on talent development and organizational effectiveness, enabling leaders to achieve positive business results.

Prior to this, Nagle was vice president of human resources at Johnson Controls for global aftermarket and original equipment groups and central functions. She also held a number of positions with increasing responsibility at Johnson Controls.

She spent the first 17 years of her career at Ford Motor Company with a variety of experiences in employee and health care benefits, labor relations, and leadership development.

Nagle also served as an adjunct professor for the College of Business at Marquette University since 2017.

Nagle earned a bachelor's degree in business administration/human resources from The Ohio State University (1989) and holds a master of science administration degree in human resources from the Central Michigan University (1997).

Education and Lifelong Learning

How important is it for companies to create lifelong learning opportunities for their employees?

A culture of learning and growth (continuous improvement) is created and sustained by the behaviors and practices of leaders

I think it is important for leaders to create conditions that enable learning and growth. In essence, they must create safe environments where it is acceptable for people to make and learn from mistakes.

These work environments need to be places where constructive feedback is sought and delivered in the normal course of business, not as an exception.

The most powerful way for leaders to create an environment of learning is by sharing their own learning goals and objectives. This includes being open about areas where they themselves are trying to develop.

Work-Life Integration

What do you think about the Lean In concept?

For me, and for most women that I am around, there is a sense of fatigue of people (articles, books, etc.) who try to tell women what they should be, how they should behave, what they should say (or shouldn't say), how they should dress, etc.

The inequality that women face (and this includes people of color, LGBTQ, and others) in the workplace today can only be addressed if we stop telling women to be more or less of something, and begin to acknowledge that it is the inequity in our institutions, not women, that is the problem.

We have to expect, and to continue to work hard to ensure, that our organizations continue to evolve until equality is achieved for every person and group.

Mentor and Sponsor Relationships

How important have mentors and/or sponsors been in your own career? Have they been men or women?

I've never had a formal mentor. I have nothing against formal mentors. They work for many people. For me, it has been more natural to find mentors who were (are) around me each day.

In this way, I have had dozens of mentors in my career, men and women. They have been my bosses, colleagues, teammates, peers, and others who have coached me and lifted me up when I needed it.

In fact, over the years, my belief system – who I am today, how I think, and how I problem solve – has been shaped and reshaped by the impressive, smart, and capable people who cared about me, guided me, and mentored me. I hope that they can be proud of me today.

These people challenged me and put me in situations that I wasn't always ready for. But they were always there at the right time to pick me up, dust me off, and push me back out there.

My advice is to seek out mentors among the people in your life who make you think differently, those you admire and respect and those who will tell you truth.

Feedback is a currency in today's workplace and you want mentors who will provide feedback with only good and pure intentions.

Taking Charge of Your Career Path

How do you evaluate when to take career risks?

I've learned over the years to focus less on the risk (or perceived risk) and more on how I make important decisions in my life and career work.

I have had plenty of failures during the course of my career.

An important lesson I've learned is that the only thing worse than failing is not correcting an error, not taking responsibility, and not learning and improving.

Every important decision will have its ups and downs. Not every decision works out in the way we think it will. Our work world and careers are just so complex.

So, to succeed, we have to develop a competency that enables us to figure out how to make new opportunities work with everything else in our lives.

Resilience

Do you consider yourself resilient? If so, how did you become that way?

A key to becoming more resilient is to be committed to being proactive.

This means always looking around the corner and being prepared for what's coming next. It's critically important for us to know how to react to and put out fires. But the real value in business, I believe, is created when smart, committed, proactive people take the right risks and make the right bets.

I do think I am resilient.

For as long as I can remember, I've had a goal and future orientation. These things change and evolve overtime.

I also think about my goals and the future regularly. I think the fact that I do this naturally focuses my mind, my will, my intentions, and helps me to be resilient.

Personal Satisfaction

What would you say gives you the most satisfaction in your career?

I come to work every day knowing that I am here to serve, and not the other way around. I believe a big part of my job is to serve the members of my team, and to serve the company.

I have the privilege to work with the smartest and most passionate people on this planet.

I tell my team that we have an opportunity to be in the front lines of transformation. I ask each of them to choose to be on that front line demonstrating their leadership and resilience, and passion.

We each have an amazing opportunity to work in an exciting, dynamic industry full of innovation and competition; to serve on teams with incredibly talented colleagues; and to learn from each other and to build friendships with people from all around the world.

These are all sources of satisfaction for me.

23

Barbara J. Pilarski

Head of Business Development
FCA-North America

Barbara J. Pilarski was re-named head of business development, FCA-North America in March 2019, responsible for negotiating and executing strategic partnership arrangements. Pilarski previously served as head of human resources, FCA-North America.

Pilarski was originally named head of business development, FCA-North America in 2009 and she served in that role until 2017. She also served as executive director-mergers and acquisitions, North and South America, Chrysler LLC. In this position, Pilarski was responsible for all aspects of merger, acquisition, and divestiture activities affecting North and South America for Chrysler, and formerly DaimlerChrysler.

Pilarski joined the company in 1985 as a financial analyst and has held various positions within the finance organization, including positions in controlling and treasury.

In 2015, Pilarski was named to the *Automotive News* list of the 100 Leading Women in the North American auto industry, an honor she also earned in 2010. Also in 2015, Pilarski was one of eight women selected by readers of dBusiness magazine for inclusion in the publication's "Powered By Women" feature, recognizing the extraordinary contribution of women business leaders in Southeastern Michigan. In 2013, she received the "Spirit of Leadership Award" from Women's Automotive Association International.

Pilarski serves on the Finance Committee of the Board of Directors for Beaumont Health, Michigan's largest health care system. She also serves as a board member for the Metro Detroit Youth Clubs, as well as a member of the Campaign Cabinet for United Way of Southeastern Michigan.

Pilarski earned a bachelor's degree in finance from Wayne State University (1985) and holds a master of business administration degree from the University of Michigan (1988).

Education and Lifelong Learning

How did you learn emotional intelligence and how has this helped you be successful?

I had a great advantage in spending nearly 20 years of my career in business development. In order to be successful in that role, I was forced to develop a critical skill: the ability to step into the shoes of the person sitting across the negotiating table from me in order to better understand his or her issues and concerns. I found this to be a necessary step in closing business transactions.

If you can figure out what someone really needs to get the deal done, then you will be in a much better position to propose a solution that adequately addresses that need. That process and experience of learning to step into the shoes of others helped in the development of my emotional intelligence.

The ability to do this is important in any role that you take on in a business environment. Be empathetic. Strive to see things from the other person's vantage point. Consistently structure resolutions to business issues and problems by looking at them holistically and not just based on what's important to you. If you can push yourself to do these things consistently, you'll be more likely to succeed.

Work-Life Integration

How have you dealt with work-life integration in your own career?

Having a career at FCA and a family are both very important to me. So, when I think about what has helped me to integrate my personal and professional life, two strategies that worked for me come to mind.

The first is to make the right choices. In effect, this means taking on the right job at the right time in your life and not setting yourself up for failure. For instance, when my kids were young and I needed more flexibility at home, I made sure that I took on jobs where I could achieve that flexibility, even if that meant passing up better opportunities at work. I didn't set myself up for failure either at home or at the office.

The other strategy I call my "a piece of mind is worth every penny" strategy—and that is that I always paid my childcare providers well. In fact, I think at some points in my career, I paid my nanny more money than I was actually taking home myself at the end of each week.

This allowed me to select a childcare provider that I really trusted, someone who was extremely reliable. This approach gave me the peace of mind when I wasn't with my children that I knew they were being well cared for, allowing me to disconnect from my family life and focus on my work life.

Mentor and Sponsorship Relationships

How did you find mentors and/or sponsors?

I'm the kind of person who keeps her head down and focuses on the work. The result of this approach was that the right mentors found me.

John Stellman, who was the vice president of mergers and acquisitions when I came into that role at Chrysler, was one such mentor who made all the difference for me.

He saw potential in me that I didn't see in myself at the time. And he nurtured a level of confidence that enabled me to take on bigger roles within the company with more responsibility and more risk—while at the same time, providing a safety net to catch me when I made mistakes or experienced failures.

This was important to me as a young female professional because I had a lot of self-doubt that I was not going to be good enough, and that I was going to disappoint.

So, I think we need mentors—in particular men—to push us and tell us no, those fears are not real. You have everything you need. You are ready to take on this work and I'm going to support you.

So I didn't look for a mentor. One found me. He turned out to be one of the most important enablers in my career helping me make my journey through this company and industry. Sergio Marchionne also served as a similar mentor, pushing me to levels I never imagined I could achieve.

Taking Charge of Your Career Path

When is it wise to listen to the fear you have of a new job or promotion and when should you ignore it?

For the most part, you should ignore that little voice of doubt. This is especially true for women because we can be a little too risk adverse when it comes to our careers. In this respect, I suggest that you invoke the 80-20 rule.

That little voice will always try to convince you to just wait and spend a little more time preparing for the role. But I'm telling you, if you have the feeling in your heart that you are ready, you probably are. And, more importantly, even if you're not fully ready, you are probably smart enough and skilled enough to figure it out once you take that position.

In considering my decision to accept my previous role as head of human resources at FCA-North America, I intentionally did not do a lot of due diligence. I knew that if I did, I might discover something that would prevent me from wanting to move forward.

I'll admit when I came into that role it was tough. It continued to be tough. But I'm a better person because of it and I'm a better leader. I have no regrets.

So I would say 80% of the time ignore the voice and take the risk.

However, you should listen to that voice 20% of the time, especially if you have personal matters to attend to at home. Your work life and your home life need to make sense together. And, if taking on a responsibility at work is going to do material harm to that other part of your life, then nothing is going to be right. So just don't do it.

Resilience

The automotive industry requires resilience, especially with the cyclical nature and demands. What gives you the internal fortitude to keep going?

My resilience comes from two places: my upbringing and my commitment to deliver for the team.

I came from a family structure where every day my parents expected me to go to school whether I had a sore throat or a cough or whatever. My dad never missed a day of work at GM. And neither did my mom. So a big part of my resilience is at the DNA level.

In addition, in the role that I have now, I appreciate that I'm part of a team and that every member of the team needs to contribute in a meaningful way in order for the team to achieve the results that it needs to sustain success.

So, I try to get a decent night's sleep every night and I come back into work swinging every single day for a greater goal and for a greater good—the team.

Being part of the leadership at FCA, it makes me feel good that people can rely on me to deliver. I never, ever want to let my team down.

And so, like I said, I come into work every day swinging, drawing on the resilience instilled in me by my parents, because I realize that it's not just about me, it's about all of the employees that we have here at FCA and the future of this great company.

Personal Satisfaction

What would others say about you or what would you like others to say about you and your influence?

When I eventually retire from FCA, I hope that people remember me as being a hard worker, someone who is tough but fair, and incredibly loyal to the company. But, more importantly, I hope I am remembered as being a true advocate for people.

I should note, at one level, that I am a very demanding person. I demand that people think deeply through issues in a detailed, disciplined way and that they reach solutions that are reasonable and actionable. I demand that people deliver on their promises and I push every day to make sure that happens.

At another level, I have a strong maternal instinct. I was once accused of being too maternalistic at work as if that was a negative quality. But I view this as a significantly positive quality for me.

So while there may be a need to be tough with my staff and others at times, to push them to be better and more focused, I'm also protecting them in two ways: first, by helping to develop their skills and capabilities and to grow them as leaders in order to facilitate more impactful and sustainable careers and second, by helping to develop the confidence and resilience they need to endure the challenges of bringing important ideas forward throughout the company.

It's my obligation to push them, to develop them, and to watch over them. My own personal version of tough love.

24

Sonia Rief

Vice President, Program Management Office
Nissan North America

S onia Rief is currently vice president, program management office for Nissan North America, a position to which she was appointed in 2019. In this role, she oversees the vehicle line profitability for all Nissan and Infiniti models sold in North America.

Prior to her current role, Sonia spent the majority of her career in R&D holding a variety of positions including a one-year assignment in Japan supporting the Renault-Nissan Alliance. Her career at Nissan started in 2001 as an engineer in vehicle program management, progressing through positions in both design and test engineering, finally becoming regional chief engineer for Rogue before transitioning in 2018 to the program management office as regional program director for midsize sedans and compact through midsize SUVs.

Sonia began her automotive career with General Motors as a durability test engineer. She holds a bachelor's degree in mechanical engineering from North Carolina State University and an MBA from the University of Michigan.

Education and Lifelong Learning

How important is an MBA or other graduate degree?

An advanced degree is less important than the demonstrated commitment to continued and applicable learning. Learning for the sake of learning is great for personal growth but less so for professional growth. For some of us, internal or external training, on the job experience, or self-teaching from available resources can be just as effective as a degree, while being more efficient in terms of time and cost. For others, the depth and breadth of learning desired is best served through a formal degree program. The decision depends on the end goal. If the goal is advancement within your current job function, specialized training might be best, and the effort and application will be recognized by management. However, if your intention is to compete for a position outside the current organization, having an advanced degree is certainly going to be more beneficial and easier to communicate than a list of classes and activities.

Personally, I chose an MBA program because I wanted to be prepared for a wide range of opportunities. Looking out into the organization at jobs that interested me, it seemed the skill set I most needed to develop was business and financial acumen. The only way to develop the breadth of understanding I wanted was going to be through an organized program. I am often asked by staff for my opinion on whether to pursue an advanced degree, the type of program they should choose, and when is the right time. And always, my response is "it depends." It depends on your goals, the expectations of your company, and your personal situation in terms of time and finances. An advanced degree is a great accomplishment and should significantly enhance or broaden your capabilities. But, it's not the degree that's a must for career advancement, it's the application of lifelong learning.

Work-Life Integration

Is work-life integration more possible at some career levels than others?

Integration of work and life is a must at all career levels, but the relative amounts and flexibility of each change as job function and level change. At any level, the integration of work and life is primarily determined by the nature of the job. Project-oriented jobs with deliverables over longer time periods allow for more flexibility in integration, though not necessarily a better balance of time, than task-oriented, short-term focused roles. Flexibility here means the ability to shift working hours, location, and interim deadlines. The absolute time required might be high but there is more opportunity to fit work in and around other priorities. Task-oriented roles, on the other hand, may require less time in total but rarely allow for flexibility in when and where the task can be completed.

As career level increases, responsibility to the company and the people in it increases. The balance between work and personal naturally shifts to accommodate a growing professional responsibility. Integration is still possible at these levels, but well-defined prioritization between personal and professional goals is critical and sacrifice of lower priority desires is usually required to keep healthy integration. However, as healthy is a completely individual definition, it is really impossible to answer universally if integration is more or less possible at certain career levels. It depends so much on what balance looks like for you at each stage of your life and your career.

Personally, I've found more flexibility as my career has progressed but have had to sacrifice a little more of my personal life at each step. Fortunately, having clear priorities about my family responsibilities and communicating them honestly within the company has allowed me to move and progress at a rate that aligns with the balance I need at that time. Strong self-awareness and acceptance of consequences is a must to keep this right level of integration. For example, I can't say I'm prioritizing specific personal goals and then be disappointed at not being chosen for a new role that would conflict with those goals. Nor can I expect the company and people I work with to sacrifice their priorities for mine. The integration of work and life not only impact us as individuals, it impacts the community around us. Every time the conditions of one or the other change, we must reassess the balance and trade-offs to keep in alignment with the life experience we're striving for.

Mentor and Sponsor Relationships

What does it take to be a good mentee?

An ideal mentee is curious, honest, respectful, self-aware, prepared, and open to feedback. Good mentees have a clear objective for the mentorship whether it be career planning, specific skill improvement, or just an improved understanding of the business. Setting an objective allows them to choose the right mentor and to set an appropriate schedule and forum: how often, how long, in person or virtual, etc. Once expectations are set, it's on the mentee to follow through by setting the meetings, communicating intended topics or even specific questions before each meeting, and arriving prepared for the discussion. To gain the most from the mentorship, in addition to good preparation, a mentee must have an open mind. Feedback or opinions of a mentor may be unexpected or not aligned with those of the mentee. A good mentee will be able to take this feedback and learn from it, using it to reframe their own expectations.

The common thread in the best mentorships I've participated in has been clearly set discussion topics though not all have been a two-way discussion. One of the best experiences I've had as a mentee was with a mentor who read my discussion topics and questions in advance and then shared his insights and opinions in our meetings. I just listened and learned. On the other hand, as a mentor, I'm most comfortable with a very interactive discussion. Mentees must learn to adjust their approach to fit the mentor's style and comfort zone. A mentor who feels comfortable and relaxed will be much more open and forthcoming than one who is not.

Overall, a mentorship is a professional relationship. A good mentee approaches it with the same or more respect, preparation, and professionalism as any other working relationship or project.

Taking Charge of Your Career Path

What intentional decisions have you made about your career and were there opportunities you received that you had never considered?

At the beginning of my career, most of the intentional decisions I made were about what not to do rather than what to do. I set boundaries on what I was unwilling to do at that time (see work-life balance above!) and then shared my interests and goals but not necessarily exact roles or positions.

Taking the approach in my work life that except for a few certain boundaries, I'm willing to try anything and to try it with a positive, energetic attitude, I've had so many experiences I would have never identified by myself. In each of these experiences, I have always come away with some benefit, personal or professional, though I didn't always see it in the moment! And often they led to a network or skill that later opened the door for another opportunity I never would have had otherwise. Looking back now, some of the most influential experiences in my career were unexpected and in areas I knew little about beforehand.

However, as my career has progressed and experiences broadened, I have become more intentional. With a deep understanding of the company and key roles, I more often express an interest in specific jobs as I also work to build the necessary network and skills for those jobs. That said, as much as I may plan, I am continually surprised and delighted with opportunities to do things I've never considered.

Resilience

How do you build resilience in a team?

Resilience is built through preparation, practice, and recovery. Preparation is putting in place learnings and processes that are well organized and well understood in advance of needing them. In usual times, these processes may not be as critical because the team has the time and resources to accomplish the objective without clear guidelines. But, in times of high complexity or urgency, having knowledge and structure in place helps to relieve the stress of the unknown. Even when processes cannot be applied as usual, the knowledge of how and why they were developed allows teams to quickly recognize and adapt in a different circumstance.

Practicing resilience may seem like a strange concept—why add stress or pressure when not absolutely necessary? But, in creating smaller situations of pressure or urgency, where the risk of failure has a lesser impact, teams have a chance to experience success and build confidence in their abilities or to learn from mistakes without dire consequences.

Following every crisis must be a period of recovery. No team can operate at maximum capacity continually. There must be an ebb and flow. In jobs where stakes are high all the time, carefully planned rotation of members and roles is a must. There must also be a time for reflection to acknowledge and reward successes and to learn and improve for the future.

Finally, throughout the entire process, the positive support and motivation of a team leader is critical. A team leader must not only bring about a successful result, they must do it in a way that leaves the entire team confident and willing to perform again.

Personal Satisfaction

What gives you joy in your job? What causes you the most angst?

For me, joy comes from success and comradery. Moments of joy that stand out in my memory are celebrating success as a team, seeing old colleagues again on new projects, sincere thanks from a staff member, and even occasional venting and commiseration with peers. Those moments of genuine, positive connection with others coming from some shared experience in our work lives make or break a job. The tasks by themselves are rarely emotional. It's the experience around a task that creates emotion and meaning. Less frequently, there have been individual moments of joy when taking a risk or overcoming a particularly difficult challenge was successful. Those moments weren't usually visible to others but were significant to me personally.

Angst, a persistent worry or fear, on the other hand, comes not from singular moments but the constant trade-off between too much and not enough. There's always a fine line between having enough information to make a decision and wasting resources or missing opportunities. The same decision process applies to prioritizing the urgent and the important or professional versus personal success. Of course I worry about short-term issues like deadlines, objectives, and personnel. And occasionally, I have the time to contemplate the future of this industry and my company's place within it. But, those thoughts and concerns are either too close or too far to be a source of continual worry. It's the everyday implications of losing perspective and balance that I fear. The big picture that defines you is not drawn in a moment. It's the culmination of fine lines drawn over time that gradually come together. I never want to step back and look with regret at the picture I've drawn.

25

Kiersten Robinson

Chief Human Resources Officer
Group Vice President
Ford Motor Co.

Kiersten Robinson is Ford Motor Co.'s group vice president and chief human resources officer. She assumed this position in April 1, 2018.

In this role, Kiersten oversees all global people processes including talent management, workforce planning, learning and development, recruiting, diversity and inclusion, compensation and benefits, and the dealer policy board. Bill Dirksen, Ford vice president, labor affairs, also matrix reports to her.

As the senior leader and corporate officer overseeing people processes globally, Kiersten ensures the development and execution of business strategies that reflect the global business environment, customer, and market needs. She reports to Jim Hackett, Ford president and CEO.

Before being appointed to this role, Kiersten served as Ford's interim human resources leader, assuming the position in November 2017 after having served as executive director, human resources, global markets.

Kiersten joined the automaker in 1995 as a labor relations representative in Ford of Australia and quickly rose through the ranks, serving in the first of several international assignments in 1997 with the Ford of Europe manufacturing organization. Following two additional positions in Australia, Kiersten moved to Ford's headquarters in 2002 to hold a variety of roles of increasing responsibility, culminating in her appointment to vice president of human resources for Ford in Asia Pacific in 2010.

While leading the team in Asia Pacific, Kiersten further distinguished herself by directing the development and deployment of a robust talent and people resource plan for the growing operation and by establishing Ford's brand as an Employer of Choice across the 16 markets in Asia.

In 2016, Kiersten was appointed to lead human resources for the Americas, and in early 2017, her role was expanded to include global markets.

Throughout her career, Kiersten has served on a number of boards within the auto industry and the communities in which she lived. These include AutoAlliance Thailand, Concordia International School Shanghai, and Inforum.

Kiersten holds a bachelor's degree in education and liberal arts from the University of Melbourne, Australia.

She lives in Michigan with her husband and two children.

Education and Lifelong Learning

How important is it for companies to create lifelong learning opportunities for their employees?

It isn't just important, it's CRITICAL! Companies that create opportunities for lifelong learning demonstrate care for their employees by putting people first. There is a synergistic effect: when employees grow, the company grows—this type of learning environment promotes innovation, creative problem-solving, and employee engagement. The workplace is one of the richest opportunities for personal and professional growth—I believe it is our responsibility to create as many opportunities for learning as possible to help foster an employee's development and impact. One new trend that I am excited about in this space is the use of internal gig boards to help socialize projects and allow employees to explore opportunities aligned with their personal development goals. I love this notion of democratized learning where employees can select opportunities that best match their interest and ambition. Experiential learning through projects or applied workshops is also one of the most effective forms of learning. Moving to a learning by doing model tailored to the employee will be key as we think about the future of work. Another key priority for companies related to lifelong learning is the role we play in helping employees adopt a growth mindset, or the belief that effort and practice will help improve their skills over time. This shift in mindset helps employees seek out learning opportunities, test new ideas, become comfortable with "failing fast," and overcome challenges which leads to better employee and company performance. One of my favorite quotes is "make new mistakes" which is all about being a continuous learner and having a growth mindset. Leaders set the tone and by modeling curiosity, a growth mindset, and by providing employees the tools, opportunities, and resources, we can help employees succeed by attending to their needs.

Work-Life Integration

Manufacturing continues to cause a divide in gender—why is this? What can be done to support more diversity?

Despite the tremendous amount of progress that has been made, job applicant pools still tend to be largely male across the manufacturing industry. One of the contributing factors is that, on average, there are less women available in science, technology, engineering, art, and mathematics (STEAM)-related fields and even fewer entering the manufacturing sector. This challenge is both simple and complex: Simple in that we need to just do it. Complex in that we need to take a holistic and versatile approach to instigate sustained change. At Ford we are firmly committed to addressing this divide through a multifaceted approach that includes the following:

1. **Build the pipeline**. We need to change this starting with elementary and middle school. We need to hold up the great female role models in STEAM-related fields and show girls and young women just how exciting and fulfilling it is to work in manufacturing. We have made a lot of progress and have more work to do to increase the number of women in leadership roles in manufacturing today.

2. **Communicate and educate**. We need to celebrate and recognize the great contributions of women in these roles today. We are very quick to celebrate pop stars. We need this same level of recognition and celebration for girls in the classroom and in the workplace who are leading the way in STEAM and manufacturing.

3. **Visible leadership**. We need to actively develop and support the appointment of women in leadership roles within manufacturing. This means ensuring there are support measures in place including active mentorship, sponsorship, and peer support.

4. **Inclusive culture**. We are also directly engaging our employees to cocreate the solution. Sustained change efforts have a greater likelihood of success when employees have direct input in shaping the strategy. For example, our manufacturing leadership team has been meeting with the Ford Women In Manufacturing affinity group to continue to invest in actively shaping the culture. They recently expanded this to include summer interns to ensure we are optimizing their experience and continuing to shape the culture that best reflects employee needs and experiences.

5. **Flexible work practices**. There are some very real constraints in providing flexible work arrangements within a manufacturing environment. This should not, however, restrict us from providing tools and programs that allow employees to meet their personal and professional needs. I am very proud of the recent changes Ford has made to its salaried parental leave policies including providing up to 16 weeks maternity leave, breast milk storage and shipping (a real issue for women on vehicle launch teams), and flexibility to return to work on a part-time schedule while being paid full time for 1 month to help working moms and their children adjust to the change in routine.

6. **"Human-centered" experiences**. Like many companies, we have great tools, policies, and practices. Their efficacy is, however, largely dependent on how well we understand employee needs and the employee experience. One of my favorite examples of this was led by our manufacturing team who for the last two/three years have invested heavily in building empathy, trust, and curiosity through a program that creates profound and impactful employee experiences centered on what it can feel like to not be in the majority. The insights and sharing have been very emotional; lots of tears, laughter, compassion; and most importantly a very human understanding of how we can impact the work environment to make it more inclusive.

Mentor and Sponsor Relationships

What does it take to be a good mentee?

One of the qualities that is most important to being a good mentee is the art of being curious. It is important to be open to new ideas and experiences, and to learn as much as you can from your mentor—who often will have different perspectives than you. Mentors can help you practice new skills, build a greater sense of self, and grow—sometimes in ways that you did not originally anticipate. Another important aspect of the mentor-mentee relationship is the importance of learning how to give and receive feedback. Learning how to ask for focused feedback is a tool that can be used for self-improvement, and I have found that mentorship works best when there is ongoing, informal check-ins. That said, it is also important to be focused and prepared for mentoring sessions to get the most out of them. For example, I encourage mentees to be very clear about the areas they want to work on so they can solicit focused feedback. If you ask someone "what feedback do you have for me?", it is likely going to be very general, or worse they don't give you any feedback. Instead, if you focus the question on a particular area you are working on such as listening skills and ask "what feedback do you have for me on how well I listened?", you will get focused feedback in return that can really help your development.

To maximize the benefits of this relationship, it is important for mentors to create a safe environment guided by respect and trust for their mentees. All of the qualities important to being a good mentee are equally important to being a good mentor. Mentorship is a two-way street—and in my personal experience, I have gained as much, if not more, from my mentee. One of the greatest gifts I have received in my career is watching a mentee exceed even their own expectations.

Taking Charge of Your Career Path

If you changed companies, what was the compelling reason and was the move beneficial? Why?

One of the most difficult decisions you will ever make in your career will be knowing when it is the right time to move on to a different company or career. I worked for a number of companies/organizations before joining Ford. I have been blessed to have a wonderful and fulfilling career at Ford and have held several different roles in different regions and parts of the business—and I have learned a lot from each one. I have found over the years that it is time to move when one or more of three scenarios present itself:

- Great stretch assignment that makes me really uncomfortable

- Opportunity to work with an inspirational, people-focused leader who empowers and stretches the team

- Opportunity to work on a project with a compelling purpose that is aligned with my values

If you are really lucky, some opportunities have all three of these elements. If only one of these elements is present, I carefully consider my current scope of work. Ultimately, the decision to change roles is all about the opportunity to learn, grow, and have an impact. As I look back on my career, I can reflect on all of the skills I have developed through each move, which has been extremely beneficial to my growth and development—both personally and professionally.

Resilience

Is personal resilience built mostly from a person's own internal resources or outside support?

I believe that it is really a combination of both—there are individual traits that may predispose individuals towards varied levels of resiliency. However, there is strong evidence that resilience can be learned, practiced, and even strengthened. Therefore, resiliency can fluctuate—there are times when you may feel more resilient than other times, and that's okay. Resilience and vulnerability can be complementary. One way that I build resiliency in my own life is through mindfulness which I have been practicing for several years. I have found this to help me maintain perspective and focus on what is important. Another tool I use is taking time for reflection and insight. For example, when experiencing challenges/change/disruption, reflection allows me to think about how I could learn, grow, and improve. It also helps me to pause and take a moment to celebrate some of the victories—no matter how small.

Personal Satisfaction

What gives you joy in your job? What causes you the most angst?

Being in the automotive business! We are experiencing unprecedented change across the industry, particularly in how we work. These questions motivate me in my current role. What do employees need to be successful? What kind of experiences are we creating for them? The tools employees need and how they work is changing so rapidly—and we have such an important role to play in creating positive employee experiences, so they can contribute and bring their best selves to work. I believe our number one priority in human resources is to create value and help make employees' lives better. One of the things that gives me the greatest amount of joy is how we are changing the way we work. Watching team members work together to reduce bureaucracy and empower employees closest to the work to make decisions, demonstrates our commitment to working differently. There is such great work going on within the company right now, and being able to celebrate this is probably the best part of my job. What gives me angst is the desire to move faster and recognizing that disruption can be both exhilarating and daunting. We need to work together to support each other build new skills, embrace change, and create a better tomorrow.

CHAPTER 26

Marianne Schrode

Vice President, Global Industry Affairs
Dassault Systèmes

Marianne Schrode has been employed by Dassault Systèmes (DS) since 2004, but a strategic partner to DS since 1991. Marianne is vice president for global affairs. She is responsible for developing business relationships with channels of industry companies, opinion leaders, and C-level executives, and to demonstrate the value of Dassault Systèmes as the premier 3DExperience company which improves business processes while inventing a sustainable future.

She began her career as an educator but transitioned to the computer and transformation industry in 1983, enjoying work at Wang Laboratories, Digital Equipment, Electronic Data Systems Corp. (EDS), and IBM. Marianne has been in the automotive industry the majority of her career with responsibility for automotive Tier 1 suppliers and global original equipment manufacturer (OEMs). She has had positions in sales, business development, marketing, organizational leadership, and industry global affairs.

Marianne received both her bachelor of science and master's degree in administrative and organizational studies from Wayne State University, but is quick to point out that she has learned much from the people she has met in her work and personal life.

She is an executive board member of the Automotive Hall of Fame, a member of the Society of Automotive Engineers Detroit Section Global Leadership Conference Committee, involved with the Henry Ford Health System Detroit Institute of Ophthalmology, "Eyes On Design" since

2005, and was an executive board member at the Michigan Science Center from 2000 to 2017.

She resides in Grosse Pointe, Mich., with her husband Bob, and is the mother of two sons, Benjamin and Zachary.

Education and Lifelong Learning

How did you learn emotional intelligence and how has this helped you be successful?

Emotional Intelligence (EI) is the capacity of individuals to recognize their own and other people's emotions. EI studies show that the degree of connection with others contributes to the ability to get things done. People prefer to work with someone they like and trust—the likeability factor. While cognitive intelligence (CI) refers to abilities for understanding information, problem-solving, and making decisions; it turns out that who you are as a person rather than what you know is what often matters most in the end. People don't care how much you know until they know how much you care. However, both EI and CI go hand in hand.

I realized at an early age utilizing EI skills put me in a position of getting attention, but and not always the attention I sought. My youthful likeability in grade school got me "Ds" in conduct, and in high school voted class clown. Tempering this trait, having faith in myself, and being willing to put myself forward, instilled in me a "can do" attitude and created an ability for direct verbal exchange with senior members in a business environment. Those that are born extroverts, with innate warmth, a good sense of humor, and a tendency toward optimism are innately at an advantage. Generational family values of love, trust, and respect have provided me with a baseline measurement for expected outcomes with people. Understanding my own shortcomings provides me a basis for knowing where I need help. I tend to be big picture and idea based. Therefore, for successful project completion, I surround myself with those who are structured, detailed, and knowledgeable.

Work-Life Integration

Manufacturing continues to cause a divide in gender—why is this? What can be done to support more diversity?

Historically, manufacturing has tended to cause a gender divide, but this is changing with early workforce education for both boys and girls. Science, technology, engineering and mathematics (STEM) programs are supported by business and now included in most school curriculums. Hollywood is helping to support diversity. The nerd girl is no longer portrayed lonely, sad, and missing out on popularity. The smart girl is portrayed as good, worthy, and socially successful. There is only a single chromosome difference between men and women, so everyone just needs to dig in and not focus on the small stuff.

An example of manufacturing change with equal access is additive manufacturing (AM)/3D printing. This business is thriving and is revolutionizing the way we work and need to train the workforce. Hard skills required are CAD, engineering, material science, design thinking, safety, and analytics. Soft skills required are critical thinking, problem-solving, interpersonal communications, and collaboration. AM jobs are high impact and high pay. As new materials, machines, and processes evolve, so do the types of products that can be manufactured. AM is being used to print pancakes, wedding rings, a car, a home, and replacement parts for space shuttles. This is an exciting greenfield industry with an inclusive gender culture increasingly populated by people who are motivated by the work of the day.

Mentor and Sponsor Relationships

Is it important to have a structured system in a company for mentoring or should it happen synergistically or both?

Mentorship programs are necessary within a company to create an environment for employee success. These programs will help with employee adaptation and growth, and will also provide a company bench strength in developing a competitive advantage. However, you can never replace those people that informally influence you in and out of the workplace…and at times you may not ever know their names.

Personally, I have never been mentored in a structured system. These programs did not exist early in my career, but if I were entering the workforce today I would seek out a mentor for guidance in avoiding individual company pitfalls. Through the years men and women have positively influenced me. Mentoring happens through example setting, good or bad. I began working when I was 12 years old. Since that time I have watched people. I learned what to do, and as importantly what not to do. I made myself aware of how a person spoke, dressed, managed their behavior under pressure, and how they used humor to make a point. These cues provided me a course for action with challenges and served as models for personal reflection. Recently, a very savvy New Yorker I was working with described me as a person with "the approach of a New Yorker and the kindness of a Midwesterner." I took the comment as a compliment and an example of a blend of mentorship learnings.

Taking Charge of Your Career Path

If you changed companies, what was the compelling reason and was the move beneficial? Why?

I have changed companies several times during my multidecade career. Everyone has different reasons for changing companies. Most people consider company growth and advancement, personal growth and advancement, salary, compensation package, and company culture. In addition to some of these compelling reasons I sought the prospect of continuous learning, leveraging previous experiences, and roles that stretched my responsibilities. As I reflect back, my experiences in changing companies has surpassed my expectations.

My first professional job was as a special education teacher. Seven years later, I was recruited into a technology company as a software educator, after attending a women in business conference and befriending one of the conference speakers. Thereafter, I changed companies every five years, which was less common then but more common now. I remained with technology companies in a sales leadership role. Teachers and preachers make the best sales people!

Do not change companies for salary alone, but know that working at one company over long periods may stall financial advancements. One company change exceeded my financial expectations but was a professional disappointment. As a person evolves and grows reasons for changing company's change. The end result for everybody should be a company they're proud of, that stimulates them, that compensates them well, that does good work in the world. I found my place at Dassault Systèmes, whose mission is to harmonize product, nature, and life for a sustainable society. This innovative company provides me a platform for my eagerness to learn, to work with people whom I admire, and blend work with home life.

Resilience

Do you consider yourself resilient? If so, how did you become that way?

I consider myself resilient in business when dealing with business challenges. Ensuring high work standards with constant energy in task completion, being inclusive with people required for success, and not becoming acquiescent or entrenched by negative vibes will help to positively develop resilience. However, I am not resilient with business issues when a challenge becomes personal. My role requires a character of high EI. I prioritize clarifying collaborative situations to ensure success. Sometimes I think the often cited "collaborative intelligence" is well named. It is a combination of emotional intelligence and group effort. But the more collaborators on a project the more possibility for misunderstanding. Success requires professional resilience, as well as personal resilience. As country singer Reba McEntire said "Success requires three bones…a wishbone, a backbone and a funny bone."

Personal Satisfaction

What gives you joy in your job? What causes you the most angst?

One of the joys I find in my job is being able to be myself while representing my company. My company role complements my personal strengths. I have three roles in global affairs—to access, impact, and nurture our business partners and customers. Soft selling is in my DNA—demonstrating value, being persuasive, being likeable, and being persistent. Customers do business with people who are credible and who they trust.

Job angst is caused by the ongoing need to be a self-promoter. Internal company metrics are often too heavily weighted towards short-term quantitative versus longer-term qualitative achievements. Soft selling is an art and it normally takes time to build relationships prior to any progress or sale to be realized. Many technology companies are not "business development" organizationally structured; hence there is a management desire for monthly, quarterly, and annual results that are short sighted. Ensuring that there is a balance of plan that is adequate for what is good for the company, the customer, and me is the goal.

27

Christine Sitek

Chief Operating Officer, Global Connected Customer Experience
General Motors

Christine Sitek is chief operating officer for the global connected customer experience of General Motors (GM), where she has P&L responsibility for the global OnStar and connectivity business. She manages a team that handles nearly 200,000 calls per day in 24 contact center sites, across 20 million connected vehicles. Sitek drives a competitive and innovative product road map while delivering an exemplary customer experience. Her responsibilities also include launching new connected vehicle services, managing growth opportunities with key business partners, and collaborating with the sales and service teams on marketing initiatives.

Sitek started her career at GM in 1989 and rose through the ranks in global purchasing and supply chain, transitioning through various assignments of increasing responsibility. She also made a series of cross-functional moves within manufacturing engineering, manufacturing operations and quality.

Sitek holds an MBA from the University of Detroit Mercy and a BA from Michigan State University. She was twice named one of the "100 Leading Women in the North American Auto Industry" by *Automotive News*. In 2017, she was a Powered by Women honoree by *DBusiness* magazine. Sitek is a board member at the Detroit Institute of Arts and InFORUM.

Education and Lifelong Learning

How important is it for companies to create lifelong learning opportunities for their employees?

Creating lifelong learning opportunities for employees is incredibly important. It shows that a company is investing in their talent and enabling employees to grow in their careers.

Lifelong learning can mean many different things, from tuition reimbursement for educational degrees to on-the-job trainings and formal leadership courses. Challenging work experiences can also provide important opportunities to learn. Encouraging employees to take on stretch projects or assignments can be as valuable as a formal training program, since learning agility is a highly sought skill that cannot be demonstrated or learned in the classroom alone. By offering a range of lifelong learning opportunities, employees can tailor the opportunities to support their professional goals and aspirations.

GM has offered me an incredible number of learning opportunities throughout my career, and I know that the education I have pursued and training I have undergone has been pivotal in my development as a leader. For example, I obtained my MBA at night while working full time, and this was supported and funded by GM through a tuition reimbursement program. I have also participated in leadership development courses at various stages of my career. Recently, GM worked with Stanford to create a transformational leadership program. I was fortunate to have been exposed to many of the Stanford professors and participate in the Institute of Design, or "D" School experience. The education and training I have participated in over the course of my career have been instrumental in my professional development.

Work-Life Integration

How have you dealt with work-life integration in your own career?

Calling it work-life integration, instead of balance, is transformational. Balance implies a 50/50 split, and when this is not attained, women can have a lot of guilt. Integration is the ability to bring smaller components together into a broader system. For me, that broader system includes my husband and my family. Work-life integration is possible because I have their support.

To achieve the best work-life integration for me, I needed to become efficient in all aspects of my life to generate more time for either my family or my career. In my personal life I adopted what I call an *outsource what I could afford* philosophy. This meant different things over the years—from outsourcing the cleaning of my house, having someone prepare meals for us, to having someone cut our grass. At work, this meant empowering team members and using technology to prioritize and focus. This outsourcing frees up precious time for me to spend quality time with my family doing the activities we love.

Most importantly, be willing to ask for help and forgive yourself for not having an ideal integration all the time.

Mentor and Sponsor Relationships

How important have mentors and/or sponsors been in your own career? Have they been men or women?

Mentors and sponsors have been instrumental in my career, and their roles have been very different over the years. I have been fortunate to have both men and women guide me in my career.

From a mentor perspective I can honestly not count how many I have had. They have been peers, bosses, employees, my husband, and friends. It has been extremely valuable for me to get multiple and varying inputs. I have had mentors give me advice on a project. I have had mentors give me advice on a job opportunity. I have had mentors give me advice on how to manage a difficult exchange with an employee or a leader.

From a sponsor perspective, I can count the number of sponsors I've had on one hand. Sponsors are leaders that know me and my brand and have personally nominated me for a position. I have worked for some of these sponsors in the past, and some of them I have not. I once had an opportunity to move cross-functionally in my career. I remember asking one of my future bosses, whom I had never met, how I became a candidate. His response was that he liked what he heard about me and that I had a good reputation.

Your personal brand is important and it has an impact on your career. I have learned that it is not who you know, but who knows you. Mentors and sponsors can go a long way in helping you develop your personal brand and advocate on your behalf.

Taking Charge of Your Career Path

What intentional decisions have you made about your career and were there opportunities you received that you had never considered?

I am a business degreed person who, in 2011, had spent 21 years in GM's global purchasing and supply chain organization. My intentions and aspirations were to continue to grow and develop within that function. Then, I was offered an opportunity to work in manufacturing engineering. For someone who was not an engineer and had never worked in a manufacturing plant, I could have come up with a thousand reasons why I was not qualified.

With input and support from my mentors and family, I took the job. The job transformed me, both personally and professionally. I have never been more uncomfortable in my career than when I stepped outside my comfort zone and took that role. However, the outcome was a significant growth and development opportunity that helped open a lot of doors for me as my career progressed.

I mentor a lot of women. We tend to want to be 100 percent qualified for a position. When we think we are less than qualified, it causes us to lose confidence. Do not let lack of confidence be the reason you do not take that next assignment. Remember your strengths and skills that you will take into the assignment. You will learn, adapt, and contribute faster than you think.

Resilience

The automotive industry requires resilience especially with the cyclical nature and demands—what gives you the internal fortitude to keep going?

I have heard many people say that the automotive industry is not for the faint of heart. While this might be true, it is the challenge and excitement of the industry that keeps me motivated. I remember my dad having one wish for me. He wanted me to work for GM. This was more than 30 years ago. In his mind that meant I was set for life and could take care of myself and my family.

The industry has faced significant changes and challenges in the last 10 years. To say we have learned to pick ourselves up and dust ourselves off is an understatement. My internal fortitude comes from the incredible alignment I have with the future vision of our company. GM has embraced the change and, in my opinion, is leading the transformation of the industry. I am proud to work for a company that has a vision of a world with zero crashes, zero emissions, and zero congestion.

Change and transformation drives us to be smarter and agile, and move faster than the competition. For me personally, this creates an environment that is highly motivating. With every transformation—big or small—an improvement sits on the other side.

Personal Satisfaction

What would you say gives you the most satisfaction in your career?

I am the most satisfied in my career when I am contributing and feel valued by my team and leader. This is not about being liked, or being on a winning team. While both of those are great, I am referring to the satisfaction I get from personally contributing and making a difference.

I also get immense satisfaction from personal growth and development. I seek out tough and complex assignments where I am learning and I can contribute. Couple this with a leader that supports me and enables me to be myself, I am incredibly satisfied and engaged.

When I reflect on my career, my performance was at its best when this magic happened.

Anna Stefanopoulou

Professor, Mechanical Engineering
William Clay Ford Professor of Manufacturing
University of Michigan

P rof. Anna Stefanopoulou is the William Clay Ford professor of manufacturing and the director of the Energy Institute at the University of Michigan.

She was an assistant professor at the University of California, Santa Barbara and a technical specialist at Ford Motor Company. She is an ASME (08), an IEEE (09) and a SAE (18) fellow, an elected member of the executive committee of the ASME Dynamic Systems and Control Division and the board of governors of the IEEE Control Systems Society.

Her innovation in powertrain control technology has been recognized by multiple awards and has been documented in a book, 21 US patents, and 340 publications (7 of which have received awards) on estimation and control of internal combustion engines and electrochemical processes such as fuel cells and batteries.

She was a member of the 2016 National Research Council (NRC) committee on fuel efficient technologies and their cost-effectiveness in meeting the 2025 US national vehicle fuel economy standards. She is working now with an NRC committee on the fuel economy standards beyond 2025.

Education and Lifelong Learning

How important is it for companies to create lifelong learning opportunities for their employees?

This is very important and needs to extend to lifelong network building. When one works in a large organization, either corporation or a university, it is important to understand, or at least appreciate, all the various facets of that organization. Meeting folks from various departments and even taking short assignments in various departments is the best and most efficient way for lifelong learning.

Work-Life Integration

Is work-life integration more possible at some career levels than others?

I had my daughter when I was a full professor, and somehow, I magically managed to have her when I was scheduled for my sabbatical, so I could have a flexible schedule for more than a year. I spent a lot of time with her and somehow I managed to have a very productive and creative sabbatical. I am not sure I would have had the luxury to enjoy this phase of my family life if I was still an assistant professor worrying about tenure and other milestones.

Mentor and Sponsor Relationships

How important have mentors and/or sponsors been in your own career? Have they been men or women?

Mentors are a great asset in one's career, and I was lucky to have a few of them. They were all men and they were all three of them immigrants from the same side of the world I came from: a Greek, a Turk, and a Yugoslavian. I am still privileged to have access to their advice and support when I need it. In general, I am very frugal in asking for help, but it is a great assurance to know they are available and ready to help.

Taking Charge of Your Career Path

When do you know that you've reached a pinnacle in a job and need to move on?

When I feel I do not learn or being challenged any more.

Personal Satisfaction

What would you say gives you the most satisfaction in your career?

Educating and interacting with brilliant young women and men in my classroom.

29

Maximiliane Straub

EVP and CFO
Bosch North America

Maximiliane (Max) Straub is chief financial officer (CFO) and executive vice president of finance, controlling, and administration for Bosch in North America, a position she has held since June 1, 2010. In addition to her role as CFO, Straub is responsible for a multitude of functions, including human resource, IT, governance, compliance, and shared services.

Her leadership has shaped Bosch's North American strategy for diversity and inclusion, corporate social responsibility, and innovation. Straub is passionate about technology and innovation, guides internal startups through the Bosch Startup Platform, and regularly mentors external startups through Bosch's partnerships with various incubators and accelerators.

Straub joined Bosch in 1993 as part of a trainee program in Blaichach, Germany. She has extensive automotive and leadership experience in the areas of accounting, controlling, manufacturing, and mergers and acquisition in Germany, France, and the USA. In 2010 and 2015, Straub was recognized by *Automotive News* as one of the "Top 100 Women in the Auto Industry."

Education and Lifelong Learning

How have you structured your own approach to lifelong learning?

Curiosity drives me. I'm not the type of person who can sit still and do the status quo for very long. I constantly push myself to learn—either formally or informally—and keep growing in new areas. I may be a CFO, but I'm not reading finance books in my spare time! I read about science and psychology, and especially love thrillers. I read to learn, to challenge my way of thinking, and to stimulate my mind.

One way I shape my personal learning is though working with those around me. I created a reverse-mentoring team of Gen Y women several years ago to better understand business topics from a different viewpoint. These women have taught me so much about our company and our industry. We tackle topics related to company culture, our approach to the market, and what it's like to be a young woman working today. They appreciate the chance to learn from me, but they don't realize I learn just as much from them!

It's also been important for me to find and nurture my peer network, through industry groups such as Inforum and by serving the community through nonprofit boards. Learning from others who have faced similar workplace challenges is invigorating. A conversation may spark an idea which leads to a new perspective or solution for my business. None of us can ever know enough. Learning keeps us in the game.

Work-Life Integration

How have you dealt with work-life integration in your own career?

It's simple—family first. I have two sons and everything I do is for them. As a single mom, there was tremendous pressure from all areas of my life. No matter what position I was in, I knew my first obligation was to my boys. Was it easy? No way! But, I made it work and still do today.

Technology has been a tremendous enabler for family balance. As someone without a local family structure, technology has been my support system. I remember being one of the first to transition from a desktop to a laptop, and how thrilling it was to log in from the home office. Little did I know how quickly technology would evolve. Being able to telecommute has redefined how we approach work-life integration. I have been very fortunate to work for a company who not only believes in family values but lives it through our culture, and this is really critical.

My mother always told me, "Your desk will not keep you warm when you're old." What she meant was that no matter how much you focus on career, and all of the success you have, it's the people in life that make it worth living. It's the family around you that will care for you and be there for you. Sometimes I meet with very driven young professionals who have a plan. They can tell me exactly when they will go to graduate school, when they will marry, when they will start a family, and even their plans for promotion. But, they are so focused on their plan that they forget to live. Life will throw many curveballs your way. It's how you react, adapt, and adjust to change that will define your success.

Mentor and Sponsor Relationships

How important have mentors and/or sponsors been in your own career? Have they been men or women?

Looking back at my career, I think I was lucky enough to have a sponsor before that word was really defined as a business term. For me, a sponsor is someone who not only guides and teaches but also someone who puts their name and reputation behind you, because they believe in you. The sponsors in my career have generally been men, and also my bosses. What I've valued most about these relationships is the learning that took place.

I've found that the best mentor or sponsor relationships have three common elements. First is trust. Both parties must be willing to trust the other for the relationship to work. Second, honesty. Conversations may not always be nice, but they will be honest. And finally, a desire to grow. In the most effective mentor or sponsor partnership, you will be pushed outside of your comfort zone.

I remember when I met my most influential mentor. I had heard of him, and he had a reputation for being personable and intelligent, as well as a great leader who was well-respected. I pursued a job specifically so I could work for him. In time, he evolved from my supervisor into my sponsor, pushing me to expand my experience into areas I simply thought I wasn't qualified. That was more than 20 years ago, and I'm pleased that we still have the same rapport.

Taking Charge of Your Career Path

When is it wise to listen to the fear you have of a new job or promotion and when should you ignore it?

My philosophy is to always say yes first, and get scared later. I try to look at fear as a form of energy. I ask myself, how can I translate this fear or uncertainty into passion for the new project ahead of me?

About 20 years ago I was appointed to the position of plant manager. Here I was, a commercially-trained female going into a high-technology, automotive manufacturing facility, without any background in manufacturing or engineering. I said yes to the position without a second thought, and then began to wonder what I got myself into! There was no looking back, so I created a plan to prepare for my first 90 days. I knew I needed to learn as quickly as possible, and started there. I learned how to read drawings sitting at my kitchen table. I studied the physics of an injection pump through textbooks and articles. When I arrived at the plant, I spent as much time as possible on the shop floor, learning processes and asking questions. And most importantly, I listened. I never went into the facility pretending to know more than any individual in the building. I respected the talent and experience of the entire workforce, and this helped me to earn trust and integrate as part of the team.

I still draw on the lessons from that experience in my career today. When I begin a new position, I focus the energy from fear of the unknown into a plan for success: quickly find the right people to trust, ask questions, and listen to learn.

Resilience

What qualities make up resilience in a leader?

Leadership is all about attitude, and this is particularly true when we look at resilience. As a leader, you are being watched. Your reaction to a difficult situation will set the tone for your colleagues and team. It is in difficult times that a leader must excel, demonstrating a positive attitude and constructive approach to the task at hand.

I believe in a learning culture in which we need to accept failure and/or mistakes. A few years ago, I highly questioned the viability of one of our innovation projects. The team convinced me to continue and it was ultimately successful. That experience taught me the difference between impossible and improbable—improbable might surprise you with success, impossible leaves no hope. Failure is a normal part of business. It's in times of failure that leaders must be resilient. You must know who you are and what you stand for, and not waiver from this foundation.

In difficult situations, I tend to get very calm. I go into listening mode. It's my job as a leader to determine which vital few topics are the most critical and focus the team's energy in those areas. In these situations I also draw heavily on the values we hold as a company, making decisions which are always rooted in honesty and integrity.

In my career I have managed a great deal of restructuring. These decisions are driven by business, but in the end, they're all about the people who are impacted. I have learned so much about the power of a team in challenging times. No matter how difficult the restructuring, I tried to be fair and honest in my communication with the teams. We were a team, from the beginning until the end, and this worked because we had earned mutual respect over time.

Personal Satisfaction

What would you say gives you the most satisfaction in your career?

There is no greater satisfaction than seeing people grow in their careers. My father lived by the philosophy, "To lead is to serve." As I've grown in my career, these words have taken on an important place in my leadership style. As leaders, we have an obligation to give back to the generations who will follow. I make it a priority to engage, to mentor, to be visible, and to be vocal.

Often times I see the potential in people that they simply don't see or acknowledge. In my time as a plant manager, I recall meeting an accounting analyst shortly after I began in my new role. She was extremely diligent in her work, and I trusted her expertise from the start. I immediately saw the potential in her to excel in finance and then some, but it took time for me to convince her of her talents and ability. That was 20 years ago, and now she is a vice president with responsibilities for the Americas region. Her talent and hard work got her to where she is today; I simply nudged her in the right direction, and helped her to believe in herself.

Being a role model for women is extremely important to me. I honestly didn't realize women looked to me as a role model until 2010, when I was honored by *Automotive News* as one of the 100 Leading Women in Automotive. Suddenly women inside and outside of the organization opened up about the example I set and how the inspiration they felt from my achievements. The reaction surprised me at first, and it was very humbling. I took the feedback to heart. I feel strongly that it is part of my responsibility to use my position in leadership to drive change. I have met many talented women who carry the same commitment to being positive role models, but we need more!

I am passionate about giving back to the mobility industry. I want to inspire women to pursue careers in this fascinating area—it's the coolest place to be for anyone who wants to work in technology. Advising tech start-ups and entrepreneurs is invigorating because I can relate to the energy they have for creating new ideas and taking risk. I help them by raising different perspectives and challenging them with situations which they have not yet encountered. It's a mutual benefit, and I love it.

CHAPTER 29

30

Dhivya Suryadevara

Chief Financial Officer
General Motors

Dhivya Suryadevara was appointed executive vice president and chief financial officer of General Motors (GM), effective September 1, 2018. Suryadevara is responsible for leading the company's global financial and accounting operations.

Suryadevara was named vice president for corporate finance on July 1, 2017. She was responsible for investor relations, corporate financial planning and analysis, and special projects.

She was previously vice president of finance and Treasurer from 2015 to 2017. In those roles she oversaw capital planning, capital market activities, and worldwide banking.

Suryadevara also served as CEO and chief investment officer for GM Asset Management from 2013 to 2017. In this capacity, she was responsible for the management of business and investment activities of GM's $85 billion pension operations and was instrumental in leading the de-risking effort for the plans. She has previously held various positions within GM's finance function and outside GM at UBS Investment Bank and PricewaterhouseCoopers.

Suryadevara has an MBA from Harvard University and master's and bachelor's degrees in commerce from the University of Madras in India.

CHAPTER 30

She is a chartered accountant and a Chartered Financial Analyst (CFA) charterholder.

Suryadevara has been recognized for her career accomplishments, including Young Global Leader (World Economic Forum), 40 Under 40 (*Fortune* magazine and Crain's *Detroit Business*), Rising Stars (*Automotive News*), and others.

Mentor and Sponsor Relationships

How important have mentors and/or sponsors been in your own career? Have they been men or women?

I have been lucky to have many people, both men and women, who have supported me throughout my career. I look for mentors with a variety of strengths so I can benefit from their best thinking and unique perspectives. Often, I find that the best ideas come from those who may think differently and have diverse points of view.

I don't believe anyone should seek out only one mentor—or stress about finding the perfect person—there are traits that you can pick up from many individuals and adapt to your own style. The key is to find what works for you from a personal-development perspective.

Also, you should not assume that people are too busy to meet with you—mentorship can be beneficial for the mentor too. Oftentimes, people truly enjoy engaging with mentees and find the relationship gratifying. Take the first step by reaching out and don't get discouraged if it doesn't work out. At worst you tried and you can always try again.

The key is to find a few people you truly click with.

CHAPTER 31

Kristen Tabar

Vice President
Toyota Motor North America

Kristen Tabar is the vice president of corporate quality at Toyota Motor North American Research (TMNA) headquarters in Plano, Texas. She is responsible for development, deployment, and management of the quality systems and policies of North American operations. Kristen directs the strategy to redefine quality in terms of customer satisfaction as well as reliability and dependability. She is passionate about women's leadership and supports various women in science, technology, engineering and mathematics (STEM) initiatives. She also serves on internal diversity advisory boards, audit committees, and investment councils.

Kristen's prior role was as vice president of the Technical Strategy Planning Office at TMNA Research & Development (R&D) based in York Township, Michigan. In this position, she was responsible for Toyota's R&D work strategies, human resource planning and management, financial resource planning and management, and all capital investments. Kristen also served as safety director and diversity champion for TMNA R&D.

Previously, Kristen served as vice president of electrical systems engineering with development responsibilities in electrical systems, products, and services. She was responsible for electronics planning, design, and development of the unique North American Toyota vehicles – Avalon, Sienna, Venza, Tundra, and Tacoma. Kristen joined Toyota in 1992 as an engineer, responsible for vehicle audio component development.

Kristen earned a bachelor's degree in electrical engineering from the University of Michigan. She is a member of the university's alumnae association, SAE, IEEE, and supports working groups for ITS, vehicle data privacy, and quality leadership. During her tenure in Michigan she served as a member of the Detroit Chamber of Commerce and Toyota representative to American Mobility Center.

Kristen is married to husband, Dan, for 24 years and has three future engineering daughters: Andrea (20), Danielle (17), and Natalie (14).

Education and Lifelong Learning

How important is it for companies to create lifelong learning opportunities for their employees?

Learning can take many forms—formal education both internally and externally, new projects, new job assignments, new business ventures, new teams, new technology, or even just new problems to solve. For companies to keep team members engaged, they must provide opportunities for meaningful work and self-development. It is critical to support team members' desires to grow and hone their knowledge. This creates an environment of innovation and accomplishment that enables them to contribute more and solve more complex problems. Proactive and regular discussions between the company and team members reinforce the importance of lifelong learning and shows respect for the team member's capability and contribution. In turn, it also helps the company form institutional knowledge, contributing to better processes, efficiency, and more sustainability.

Work-Life Integration

Technology has enabled the move from work-life balance to work-life integration. Do you think that the pendulum will swing back to more separation?

Technology is a set of tools that help team members be more efficient and effective. The boundary between integration and separation is up to the individual and the career or business they choose. There are times when integration is necessary and there are also times when separation is necessary. These are not mutually exclusive. We are lucky there are technologies that give us the flexibility to decide where on the spectrum we need to be and change when we want to. I think the pendulum is always moving—it's a continuum based on work needs and life needs. Most people are constantly assessing how to shift and prioritize to maximize both their work and their life experiences.

Mentor and Sponsor Relationships

Is it important to have a structured system in a company for mentoring or should it happen synergistically or both?

Mentoring is a career-long journey. You always need a lot of advice and perspectives to inspire you to create different solutions. Many mentee-mentor relationships develop organically based on people you work with directly. These may seem natural since there is usually a common project or problem that initiates the mentorship. Over time, trust is built, and exchange of knowledge grows. However, team members know the benefit of adding a more company-structured program that pairs team members with a certain skill or knowledge with another team member seeking to learn and improve in that area. These two team members may not have an obvious project or problem that allows them to work together. However, using internal networks and structured mentor programs enables relationships to get started. These structured programs are a valuable way for team members to develop their own internal networks and broaden their understanding of other parts of the company. They also provide good visibility to talent across the organization.

Taking Charge of Your Career Path

What is the best strategy for asking your company for a new assignment?

The key to asking for a new assignment is knowing what you want and why, and grasping if you have, or if you can attain the skill set for the next step. You need to spend time thinking about yourself, the work that motivates you (and the work that doesn't), what type of things you want to learn, where you want to contribute, and what parts of your company can give you those opportunities. How does this build your capability or leverage your strengths? How does this prepare you for the next assignment? Talk to your mentors and people in your networks that work in those parts of the company. In other words, do your homework before it's due. Then set a discussion with your supervisor, explain what you've been thinking, and clearly declare the new assignment you want. This is not an ultimatum but a discussion about your passion, growth, and value. You need to be clear about what you want and why it will benefit you and the company.

Resilience

The automotive industry requires resilience especially with the cyclical nature and demands—what gives you the internal fortitude to keep going?

If we look at the automotive industry very near term, the cyclical nature and demands can feel uncomfortable. However, that's short-term thinking. The automotive industry is experiencing a transformation. We are rethinking our business models and products. Everything is changing. The types of products are expanding and crossing over to other forms of mobility-personal, shared, mass-transits, alternative propulsion systems, robotics, avionics—and more. Technology is integrating to our products and services at an exponential pace. There is no limit to where this will take us. This unimaginable opportunity gives us the inspiration to keep thinking about our customers—how do we create the things they want and help them realize the things they can't yet imagine. These are the thoughts that see us through the cycles and demands of today and help us look to tomorrow.

Personal Satisfaction

What gives you joy in your job? What causes you the most angst?

I'm a people person who loves to solve problems. The automotive industry gives me the unique opportunity to solve problems and create new things that are used by literally millions of people every day. While we may not be IT start-up fast—yet, the automotive industry has the unique situation where you can dream of something new, design it, test it, build it, sell it, service it, and within a short period of time "it" can become that beloved first drive or that quiet "alone" time or that exhilarating getaway or someplace to transport you and your family where they want to go. The problems we solve and the things we make in the automotive industry are for people. Connecting with them and making their mobility better gives me great joy.

Lynn Tyson

Executive Director, Investor Relations
Ford Motor Co.

L ynn's career spans over 25 years of leadership roles in treasury, international corporate finance, corporate communications, and investor relations (IR) serving in senior level positions at Fortune 100 companies. Lynn has also provided strategic communications consulting services to companies looking to enhance the effectiveness of their investor relations (IR) programs. Specialty areas include IR as a competitive advantage, strategic communications, IPOs, and crisis communications. Lynn's capabilities in leading world-class IR functions have been recognized by the equity market across several sectors, and she is also a recipient of the Silver Anvil Award for excellence in crisis communications from the Public Relations Society of America.

Lynn is currently executive director, IR for Ford Motor Co., where she is responsible for leading all of Ford's IR initiatives, including representing the company to equity and fixed income investors and rating agencies, and providing strategic counsel to support value creation and risk mitigation. Lynn's career includes 14 years of finance and strategic communications experience with PepsiCo, where she ultimately served as senior vice president, IR, and 10 years at Dell, where she led IR and global corporate communications. Prior to Dell, Lynn led IR for YUM! Brands, where she was involved in its spin-off from PepsiCo in 1997.

Lynn is a native New Yorker and an avid equestrian. Lynn holds a bachelor's degree in psychology from the City College of New York and an MBA in finance and international business from the Stern School of Business at New York University.

Education and Lifelong Learning

How have you structured your own approach to lifelong learning?

I was raised to embrace life as a learning laboratory—to listen; to observe; to be open-minded, inquisitive, and fact-based; and that there are no limits on learning or knowledge, including time. Also, that it was important to be self-aware of what you know and what you do not know—and not to be afraid to ask for help when you need it. In addition, learning is not necessarily defined by or limited to scholastic achievement. I grew up in New York City where being "street smart" was essential. I did not learn that in school—I learned it from my environment, including my family, and I've been able to apply those street smarts to all aspects of my life, including business.

Work-Life Integration

Have you ever chosen work over family or vice versa? Has this gone well? What did you learn?

I do not think work and life are binary—it is a false choice. For me the more important question is "What kind of a life do you want to lead?" Then you shape that life based on the decisions you make and the people you have in your life—especially your life partner. Being happy, nurturing a strong family, indulging in my passion for horses, and being of service to others—that is the life I chose to lead. Work is not my life nor is it who I am. To be clear—I love my career and yes, over the years, I have occasionally missed parent-teacher conferences, birthdays, and anniversaries. However, to me that is not about choosing one thing over another—it's a known requirement of my job that my family and I embrace. What is far more important to us is the kind of life we lead.

Mentor and Sponsor Relationships

How important have mentors and/or sponsors been in your own career? Have they been men or women?

Mentors have been incredibly important to me. They have taken many forms: family, friends, people senior to me at work, peers, and people at lower levels than me. Anyone who has experienced something I have not knows something I do not, or sees something in me that I do not appreciate (bad or good), I view as a potential mentor. Very early in my career a handful of people, mostly men simply because of the era, extended themselves to me, educated me on how corporate America works, how to navigate, and what to expect as I moved up the ladder. With each career move, I sought out people I could learn from.

Taking Charge of Your Career Path

What intentional decisions have you made about your career and were there opportunities you received that you had never considered?

My career path has been circuitous—and this has taught me that I don't always have the answer and that I need to be open to change. Growing up I wanted to be an equine veterinarian. So that is the path I initially took in college. But this eventually changed and I wound up with an MBA. I enjoyed business and I knew a career in business could fund my passion for horses.

Once in business I thought I wanted to be a general manager—owning a P&L—and I was tracking in that direction. Then the CEO of my company suggested I do a rotation in investor relations—something I knew nothing about. I did the rotation assuming I would eventually resume my general manager track. I found that I loved investor relations—it capitalized on my strengths while requiring work on my developmental areas. I embarked on a new career and opportunities unfolded—in different countries and in different sectors. At another point in time, yet another CEO challenged me to broaden myself—and lead a global communications function in addition to investor relations. This was something I had never considered but he thought I was up to the task even though at the time I was skeptical.

Resilience

Is personal resilience built mostly from a person's own internal resource or outside support?

Resilience has to come from within—it must be ready in an instant to overcome challenges and persevere even when people tell you that you cannot. Others can nurture it, such as family and friends — but I think you must own your own resilience. Resilience for me is a combination of strength, confidence, capability, and courage of conviction.

Personal Satisfaction

What would you say gives you the most satisfaction in your career?

Being very good at what I do and developing others (especially as I have gotten older) gives me tremendous satisfaction.

33

Carrie Uhl

Vice President, Purchasing and Supply Chain, the Americas
Magna International

Carrie Uhl was named vice president of purchasing and supply chain for the Americas in July 2010. In this position, Carrie is responsible for coordinating the supply chain function across Magna International's North and South American operations and ensuring consistent and strategic approaches to supplier selection, category management, and supply chain risk evaluation.

Carrie joined Magna in 2005 as commodity manager for Magna Exteriors and Interiors, where she set strategy and negotiated agreements for plastic resin, the operating group's largest purchase category. She has progressed through positions of increasing responsibility within the purchasing function, including global commodity manager for Magna International and then also director of purchasing for Magna Composites.

Prior to joining Magna, Carrie held purchasing positions at both Lear Corp. and Guardian Industries.

Carrie earned a bachelor of science degree in human and organizational development from Vanderbilt University in Nashville, Tenn. She is a member of several automotive industry councils and the advisory board for Wayne State University's global supply chain program. She serves on the board of directors for Junior Achievement of Southeast Michigan.

In 2015, Carrie was named by *Automotive News* as one of the "100 Leading Women in the North American Auto Industry," and she was also included in its "2016 Class of Rising Stars" for automakers and suppliers.

Education and Lifelong Learning

How important is it for companies to create lifelong learning opportunities for their employees?

I think it is critical that companies create lifelong learning opportunities. It's a key piece of any good retention strategy. Promotional opportunities may not always be readily available, but people will wait if they continue to develop new skills in their current job. It signals that the company is invested in you and your future. Plus, every company should want their employees to be at the top of their game. Without opportunities to learn, employees will become stale and disengaged, and the company will fall behind its competition pretty quickly in terms of innovation and customer satisfaction.

Work-Life Integration

Technology has enabled the move from work-life balance to work-life integration. Do you think that the pendulum will swing back to more separation?

Technology is really a godsend for work-life integration. Earlier in my career, it wasn't common to have a laptop. When I had my first child, it was so difficult to be as productive as I was before I started my family. I didn't want to miss those precious couple of hours with my daughter between dinner and bedtime, so I left at 5 p.m. and lost the hours that would have been spent at the office in earlier years. Today, I can pick right back up at 9 p.m. and get a few hours of work in from my couch. Cell phones enable conference calls to be taken during a commute. Bluetooth speakers allow it to be done safely. The Facebook Messenger app lets me see my family's smiles "live" when I wake up in Shanghai and they are finished with dinner. I don't think the pendulum will swing back. Work-life integration is here to stay, and thank goodness.

Mentor and Sponsor Relationships

How important have mentors and/or sponsors been in your own career? Have they been men or women?

More than 20 years ago, Joe Bruce, the vice president of global purchasing for Guardian Industries at the time, took a chance on hiring me as a purchasing agent, with no real experience. In the years that I worked for Joe, he encouraged me to present my own work and ideas to various company executives, even if I was worried I was too junior or didn't feel ready. "It's your work," he'd say. If I hesitated, his encouragement was "You'll be fine. And if you make a mistake this time, just don't make it twice." I grew leaps and bounds in that environment. He was not only a mentor, but also a sponsor, opening doors to new opportunities to learn and grow.

So 15 years later, when I was chosen for the position I have today with Magna, I called Joe. We had stayed in touch over the years and I wanted to share my good news, but also my nervousness that maybe I wasn't 100% ready for this new task. As usual, Joe reassured me. And then three days later I got a card in the mail.

It is one of those Successories cards, and the heading says Achievement. Underneath, it says "Unless you try to do something beyond what you have already mastered, you will never grow."

Most of us lug around a heavy laptop bag. In mine today, you would find an umbrella, a stale protein bar, and a half dozen device chargers tangled in a big knot. But one staple in my laptop bag for 10 years has been Joe's card. I still pull this card out and look at it from time to time when I need a boost of confidence. That's evidence of mentoring that sticks.

Taking Charge of Your Career Path

If you changed companies, what was the compelling reason and was the move beneficial? Why?

Earlier in my career, I made a strategic decision to leave a company and a job that I loved. The only reason was because the organizational structure was so flat, there was not going to be room for a "traditional" promotion or career path, even though I was able to take on new challenging assignments and learn and grow. The new company was known for having many management layers, and sure enough I was able to move up the ranks. I didn't stay there long because the culture didn't suit me, largely because of those layers. But sometimes, especially early in your career when you are getting established, you do need to make sure you have a strong resume on paper. Then later in life, it becomes less about the resume and more about your network and the depth of your experience.

Resilience

What qualities make up resilience in a leader?

I think resilient leaders have a few key traits. First of all, they give themselves grace, and they give others grace. Everyone makes mistakes. If you don't make mistakes, you aren't taking enough risks. So the key is to embrace your mistakes as learning opportunities. Resilience also requires perspective: knowing that failure on a task doesn't make you a personal failure. Resilient leaders recognize that nearly all positive change happens in times of stress. I love the analogy that diamonds can only be made under pressure. I think the same is true of humans and how we get better. Finally, resilient leaders talk about their missteps. They don't hide them. If you cannot be a vulnerable leader, you create a culture where your employees cover things up and soon you have an avalanche instead of a snowball that could have been dealt with sooner and more easily.

Personal Satisfaction

What would you say gives you the most satisfaction in your career?

Nearly my entire career has been in purchasing and supply chain. This function in many companies used to be considered more tactical. "Buy my things for me, and get them here on time. And cheap." I have always tried to make sure purchasing and supply chain is recognized as a strategic function, and one that is essential for delivering customer needs and shareholder value. Throughout my career, a lot of the work that I do is focused on driving that transition to a truly strategic function, not just within my company but also within the broader automotive industry. Seeing that take shape is what gives me the most satisfaction.

34

Marlo Vitous

Head of Supply Chain Management Planning
FCA—North America

M arlo Vitous was named Head of Supply Chain Management planning, FCA North America in June 2019 and is responsible for the optimization of supply chain planning and operational activities.

Prior to this, Vitous was director of interior and electrical purchasing responsible for leading a global team in the development and execution of long-term and cross-functional commodity strategies.

Vitous has been with the company for more than 20 years and worked in various roles in manufacturing, supplier quality, and purchasing.

Vitous is a member of the Women's Alliance at FCA and a member of AutomotiveNEXT's executive committee at Inforum. She was also recognized as one of the *Automotive News* "100 Leading Women in the Auto Industry" in 2015.

Vitous earned a bachelor's degree in science and business administration from Central Michigan University (1998) and a master of business administration from Wayne State University (2001). She is a certified procurement manager (CPM).

Education and Lifelong Learning

How have you structured your own approach to lifelong learning?

I encourage you to keep taking risks. If a new job, project, or manager scares you, understand that you will probably learn and develop from that experience.

Some of my best learnings were from opportunities and people that were different from my background. Listen to and learn from feedback. The feedback I've received from past and current managers, colleagues, and employees was valuable for my development and in taking on larger responsibilities.

I've also actively taken advantage of opportunities to learn. I've:

- Always taken leadership training when offered and asked for more.

- Striven to employ the leadership tools I've learned, and have worked to teach others how to use these tools. This has sparked conversations, deepened my own learning, and helped others develop as managers and leaders. I continue to be a mentee and mentor for employees.

- Joined professional networks to enhance my external relationships.

- Read. It helps me think differently, increases my knowledge, and reinforces the ideas I have about leadership and performance.

- Kept up with technology trends to see how these can make me and the team more efficient in my job and in my life overall.

- Attended workshops that may stimulate new thoughts and increase my knowledge.

- Listened to people who have experiences that I can learn from and ask many questions.

And, finally, the automotive industry is dynamic, so be curious and make learning a priority. Working at a company that supports your learning is so important for your future development as a leader.

Work-Life Integration

How have you dealt with work-life integration in your own career?

I'm living my life the best way I know how. I love having my family, friends, health, and career, while maintaining my core values.

These are priorities in my life, and it takes a lot of hard work keeping up with it all. The rewards from each of them are worth it to me, so I've learned how to prioritize at different points of time when conflicts arise.

It's all about choices.

My priorities are like balls: some are rubber (important, but they will bounce back) and others are glass (priceless, but they will shatter when you drop them).

First, you need to be comfortable with the juggling and be present while juggling. That's the hard work.

Next, you need to make sure the balls you drop are the rubber ones. Not the glass.

At times, you will do it well. Other times you will fail. You must learn from both your successes and failures prioritizing and juggling the rubber and glass balls in your life.

So how does this work for me? An example: My kids play soccer. Some games are important to my kids. Those become a priority glass ball.

I also try to be present in work meetings, so more meetings aren't required. And, I listen to my family and friends when I am with them.

I've also learned to say no to events and meetings. Life will pressure you to do everything. You need to live the life you want and set boundaries.

I believe the life you have is the one you create. I do not waste time worrying about the rubber balls I've dropped. I also try to reflect on making a difference with my family, friends, health, and career. I live life intentionally.

Mentor and Sponsor Relationship

Is it important to have a structured system in a company for mentoring or should it happen synergistically or both?

Both are important.

My friend and I started a structured mentoring program in our group at FCA because we believed the organization needed one, and we wanted to make a difference. We called it "People to People" mentoring. We believed a formal program would drive learning and support for both mentors and mentees as well as benefiting our organization by increasing employee engagement, knowledge, and retention.

Mentees have three goals and are paired with mentors based on those goals. This enabled better matches. Our committee also created a social community network by having three events per year tied to the community, speakers, and training.

The program has grown and now features peer-to-peer mentoring relationships, which enables newer employees to grow their experiences and familiarizes them with our work culture.

I believe a formal mentorship program connects employees with each other and gives them platforms to discuss their ambitions and challenges. It also creates a learning organization. I've had the pleasure of being both a mentor and mentee and still, to this day, treasure those relationships.

It's important to own the development of future leaders. Some of my best mentoring has come from managers and colleagues and I'm forever grateful for those connections. I have created a personal board of advisors based on these relationships and call on them when I need advice and guidance in my career.

My favorite mentoring relationship is from a group of ladies who share a common journey with me. We work in the same industry and are in leadership roles. We also have similar values, yet are all very different individuals. We share in each other's successes and failures and have become strong supporters of each other. I'm thankful for them every day.

I recommend finding and nurturing these types of relationships both formally and informally within your company.

Taking Charge of Your Career Path

What was the biggest surprise in your career when you faced a new opportunity thinking it was a terrible move, but tackling it in any case – what did you learn and how it change your view?

It was early in my career at FCA. I had worked in manufacturing and supply chain and was looking to move to purchasing for another cross-functional experience.

Based on advice I had received, my sights were set on becoming a seat buyer. Seats are a great commodity, and it was a good team to be on. Twice, I interviewed, and twice I was rejected. So I waited.

One day, a manager in the steel purchasing area said he would interview me but that he also had many qualified candidates.

Becoming a steel buyer was not on my dream list of jobs, but I told him I would be the best buyer he had ever seen if he gave me a shot. I got the job.

The steel market became very challenging that year, so that job became a critical role. The experience changed my viewpoint in so many ways.

I learned to be humble – always. Sometimes, you need to get yourself in the door first when looking into a new area, and your first job choice may not be the one you get. This job forced me to learn quickly and to be flexible and adaptable. I was able to show my strengths and received a lot of visibility in front of senior leadership. This group also needed my skills and I was able to bring value to that team right away.

I've learned that there is no one ideal career path to follow and there is no certainty in what we do. Just continuously learn, grow, and be of value to the team with outstanding performance. A great career path will follow.

And here's the irony: I never became a seat buyer, but became the global head of interior purchasing, which includes seats.

Resilience

Do you consider yourself resilient? If so, how did you become that way?

Yes, I'm resilient and I've become this way by adapting to the many changes, challenges, and opportunities at my company (Chrysler Corp., DaimlerChrysler, Fiat Chrysler Automobiles), and in my own personal struggles. Having challenging goals means you will experience obstacles that you may not be able to overcome without having the passion and perseverance to succeed.

I've been able to grow and thrive in my career during those challenges and changes. I see it as having grit and a positive attitude, and as said nicely by one of my team members: "We've got this."

Resilience is important for employees at work and in life. We live in a world of constant change and adversity.

Some tips:

- Develop short-term and long-term goals.
- Celebrate the small wins.
- Manage your thoughts and emotions in stressful situations.
- Visualize your success.

Also, have high-quality relationships and connections within your company and family. A social network will support you in times of doubt. Note that it takes energy to give energy. We need self-care and recovery. When my energy starts to run low I know I need to take the time to recharge myself.

Mindset is important. You have to have mental toughness and flexibility. Be authentic in how you lead and work in accordance with your values and strengths. My work matters and brings value and impact. My commitment to my teams is that I'll never back out.

I want to help influence and control the outcome and not give up and, while it may be stressful, I see it as a challenge that I need to grow through versus allowing it to take me down.

In the end, you need to believe in yourself, your team, and practice optimism.

Respond versus react.

Pause and reflect on the issue from a neutral space.

Then, try to solve the problem versus just reacting to the circumstances.

I believe creating a compassionate work culture with resilient leaders is most productive. You need to be prepared to face any challenge in this competitive environment that constantly presents disruptive forces.

Personal Satisfaction

What would you say gives you the most satisfaction in your career?

The most satisfaction I have is when I know that what I do at work matters – when it matters to the business and the people.

More importantly, it matters knowing that I am making a difference by leading people: I am coaching, empowering, and developing leaders who have the courage to challenge the status quo, face new challenges, and create new paths.

I enjoy motivating people to bring out the best in themselves and watching them grow in their careers through guidance, support, and recognition.

I enjoy creating an environment that inspires people and contributes to a sense of community so FCA can win as one cohesive, fully empowered team.

There is always something great to achieve in the automotive industry, and encouraging others to feel that way excites me. When I leave one area for another, I hope the people and group are better off than when I started. It's that simple.

Judy Wheeler

Division Vice President, Dealer Network Development and Customer Quality
Nissan North America, Inc.

Judy Wheeler is division vice president, U.S. and Canada Dealer Network Development and Customer Quality, Nissan North America Inc., a position to which she was appointed in May 2019.

In this role, Judy is responsible for all Dealer Network Development (DND) operations in the USA and Canada for both Nissan and Infiniti brands and Customer Quality, which includes customer resource centers, all dealer and field personnel training, and customer-focused initiatives through the dealer body.

Previously, Judy was division vice president of Dealer Network Development. Prior to that she was the division vice president for sales operations and regions, Nissan North America, Inc. She led the Nissan division sales operations, vehicle operations, fleet, and remarketing functions and was accountable for all facets of Nissan's U.S. domestic sales activities with a focus on increasing revenue and profit generation for the Nissan brand and dealer network, as well as driving sales performance in the Nissan Regions.

Before this she also served as vice president, Nissan Southeast Region, Nissan North America Inc., where she was responsible for regional sales and marketing activities in Nissan's Southeast Region. Before that, she was director of marketing for Nissan Canada Inc., where she was responsible for all planning and implementation of marketing communications and media for national, retail, digital, customer relationship management, social sponsorships, and auto shows. Judy was also responsible for product planning, day-to-day marketing actions, and intelligence gathering, plus overseeing Nissan and Infiniti incentives.

Judy holds a bachelor's degree in business administration from the University of Wisconsin and a master's in business management from St. Mary's College, and she has successfully completed the Harvard Business School executive management program, as well as the executive leadership program at the Wharton School of Business. She is also the executive sponsor of the Women's Business Synergy Group at Nissan North America.

She is based in Franklin, Tenn.

Education and Lifelong Learning

How have you structured your own approach to lifelong learning?

Throughout my career, I have identified a specific area I want to improve upon professionally and then created a training plan to achieve it. I started this practice early on and obtained my MBA while working full time; 10 years later, I completed the Harvard Business School Executive Management Program and earned a degree this past year from the Executive Leadership Program at The Wharton School of Business. My strategy each year is to look for growth opportunities and programs to help me meet my development goal in areas such as leadership, coaching, mentoring, strategic thinking, and motivation of top-performing teams.

As a young professional, I needed to work on having a voice that was heard in a male-dominated environment and focused my developmental efforts on my assertiveness as a businessperson. One of the most beneficial trainings I took was a seminar on negotiations that has proved to be a valuable competency through the years.

Work-Life Integration

Have you ever chosen work over family or vice versa? Has this gone well? What did you learn?

Family First—period! My family has moved 14 times both domestically and abroad. After we had our children, I was offered a promotion in another state. My husband and I decided then that it would be best for our family if we followed my career and if he started his own company. It was a choice that would give us a better balance between family and work, and one where I would be confident that our children were getting the best care possible.

As our children got older and we relocated a few more times, I promised them we would return to the U.S. so they could complete high school at one school. I kept this promise, but it wasn't easy. My personal sacrifice was to take two U.S.-based positions that required extensive travel. In the first role, I traveled globally three weeks of each month while the second assignment required me to work in Canada during the week. From my children's perspective, this arrangement was better for them, and my husband was supportive, but it was physically exhausting for me.

Our children are now adults and they often comment how appreciative they are of the sacrifices we made to ensure they experienced as little disruption in their life as possible. My husband has been wonderful throughout the years supporting this decision. It takes a strong marriage and commitment to each other and to your family to make this work.

Mentor and Sponsor Relationships

Is it important to have a structured system in a company for mentoring or should it happen synergistically or both?

Mentorship is one of my passion points. I have made it a priority over my career to dedicate time to serve as a mentor and ensure programs exist within the companies that I have worked at to provide a formal mentorship program. I believe having a structured mentorship program is of the utmost importance, which is why I have started my own mentoring program at every location I've called home for the past two decades. You can have casual, one-off mentoring, but let me illustrate the power of a company-based program.

Each year I launch a formal mentoring program for the team that reports to me, but anyone in the organization who is interested is welcome to join. My program typically has approximately 100-160 participants. In the eight years I've been with my current company, my mentoring program has been shared with other internal business operations, including Canada, Mexico, regional sales offices, and our financial division.

For each session, I form a committee that helps with the management of the program. This is purposeful as I want the program to be sustainable and shared with other functions in the future. Each mentor and mentee must complete an application. This application is used to create matches based upon the strengths of the mentors and the opportunities for improvement or career interests of the mentee. We have a specific guide to follow to kick off the program. All participants attend a one-hour session that I personally lead, and then hold quarterly touchpoints to see how things are going. The mentor/mentee pairs meet monthly for one year. We also include a career development plan activity that is currently being uploaded in our HR system. At the end of the year, our mentors/mentees share their experiences and we have a celebration.

Additionally, I am actively involved in the Women's Business Synergy Team as an executive sponsor, which supports female employee professional growth and development. This year, we also launched a mentorship program for this group that is open to both women and men.

As one leader, I have influenced thousands of people over the past 16 years with my mentorship program. Feedback over the years is that the program has been rewarding for mentees and mentors. For me, it will be my professional legacy.

Taking Charge of Your Career Path

What intentional decisions have you made about your career and were there opportunities you received that you had never considered?

Maintaining an action plan has been instrumental in guiding my career. First, I create a timeline listing the roles I want two, four, and six years in the future. Next, I write down what I need to do to be successful in my current position and identify ways that I can exceed my leadership's expectations. Then, for each of the future desired positions,

I outline what skills I will need to have to position myself as the best choice for that role from executive leadership's viewpoint. One way I've put this plan into action is to volunteer to work on a special project for the executive who oversees the position I aspire to have in the future. Last, I check my plan regularly and update accordingly.

My action plan has prepared me to respond quickly to unexpected opportunities that arise. For example, the CEO at a former employer asked me to take a role that would push me out of my comfort zone. Although I had zero expertise in the functional area, I didn't want to say "no" to the opportunity. Instead, I went into the new position with a clear strategy to work hard and exceed expectations by combining new ideas and out-of-the-box thinking with an energy that was contagious. My goal was to get promoted quickly so I could get back into a role that I enjoyed. I earned my next promotion within 18 months. Looking back, it was my least favorite position ever and I've held 28 roles over my 35 years in this industry.

Resilience

The automotive industry requires resilience especially with the cyclical nature and demands—what gives you the internal fortitude to keep going?

My upbringing as I was raised on a dairy farm. It is a business that is influenced by the fluctuation of milk and crop prices, the weather, supporting your neighbors, and how well you manage your finances. My father is an excellent businessman who taught me so much about how to run a successful business that I will be forever grateful. He also taught me how to take risks and weigh the pros and cons on the likelihood of success or failure. I also learned the importance of teamwork and working to find solutions to sometimes overwhelming issues. We also celebrated success, and we all understood and respected the importance of family.

Over the years, I have used these same principles and lessons when working with teams, understanding what my boss requires, motivating a team, and being a truly inspirational leader. Yes, we will have good years and bad years, but I try to focus on the good each and every day.

Personal Satisfaction

What would others say about you or what would you like others to say about you and your influence?

Others would say that I am a warrior for my people. That my mentorship program has really made a difference in their personal and professional growth. Finally, they will definitely say I would do anything for my family!

36

Kate Whitefoot

Assistant Professor, Mechanical Engineering,
Engineering & Public Policy
Carnegie Mellon University

Kate S. Whitefoot is an assistant professor in the departments of mechanical engineering, and engineering and public policy at Carnegie Mellon University. Professor Kate's research bridges engineering design theory and analysis with that of economics to study the design and manufacture of new technologies and their adoption in the marketplace. Her areas of expertise include vehicle fuel efficiency, consumer choice, the design and adoption of green products, manufacturing productivity, and energy and environmental policies.

Professor Kate has gained recognition nationally and internationally for her research and teaching. She currently serves on the National Academies Committee on Assessment of Technologies for Improving Fuel Economy of Light-duty Vehicles. Her research is published in *Science*, the *Proceedings of the National Academy of Sciences*, *Environmental Science & Technology*, and the *Journal of Mechanical Design*, among others, and is featured in the *Washington Post*, *Popular Mechanics*, *Bloomberg Business*, and *Business Insider*, and referenced in the 2017-2025 Corporate Average Fuel Economy rulemaking. She has worked with several companies in automotive, aerospace, and high-tech industries and has been invited to present briefings at the White House, Capitol Hill, the Department of Commerce, and the Environmental Protection Agency.

Dr. Kate earned three degrees from the University of Michigan: a BS and MS in mechanical engineering, and a PhD in design science—a multidisciplinary program where she concentrated in mechanical engineering and economics, completing course sequences and having an advisory committee across both disciplines.

Education and Lifelong Learning

How have you structured your own approach to lifelong learning?

Lifelong learning is critical for both individuals and organizations to continue to grow and stay up to date with evolving technologies and information. I set aside a few hours every week that are specifically dedicated for me to learn new things that I might put off at other times because I'm focused on producing new research or teaching others what I know. Being disciplined about focusing this time on either broadening my understanding of new technologies or disciplines, or staying up to date with new techniques in my field has been invaluable for me to be innovative in my work.

Work-Life Integration

How have you dealt with work-life integration in your own career?

For me, the most important aspect of work-life integration is to be authentic in both my work and home life so that I am a "whole" person rather than having my career goals compete with my life goals.

Mentor and Sponsor Relationships

How important have mentors and/or sponsors been in your own career? Have they been men or women?

I have benefited enormously from very rewarding mentor relationships from both men and women. Some of these have been formal advising relationships that have lasted for many years after the formal commitment ended. And, some of my mentors have come from unexpected connections that have developed into very rewarding relationships.

Taking Charge of Your Career Path

What intentional decisions have you made about your career and were there opportunities you received that you had never considered?

My career path has been a combination of planning for success and also being flexible to take advantage of unexpected opportunities when they're presented. As a result, I've ended up taking a path that I couldn't have envisioned 15 years ago, but that still embodies my principles and ambitions. Looking back, I am very happy that I chose to be flexible in the direction my career was headed.

Resilience

What qualities make up resilience in a leader?

The quality I admire most in leaders is the ability to be open to new information and changing one's mind. Some of the strongest leaders I know are able to easily let go of previously held beliefs when faced with well-founded information that contradicts them.

Personal Satisfaction

What would you say gives you the most satisfaction in your career?

One of the most joyful parts of my career is when someone I don't know comes up to me and tells me that they've read one of my papers or books and that it influenced them. It's great to see the impact that the work has had on many different people and organizations.

final thoughts

A sustainable career—one that fits our values and is flexible enough to evolve as our interests and life stages change—is a goal that many of us have. What can we learn from the women in this book about how to achieve that goal in the automotive industry? Here are three of the not-to-be-missed themes, along with a few of our own observations.

The importance of owning our own careers.

The truth is that no one is more qualified and motivated to manage our careers than we are. We know our superpowers and our passions. And we can learn how to be effective and powerful advocates for ourselves so that others recognize and notice our capabilities and value. While we cannot predict or control what happens around us, we can position ourselves to be ready and open when opportunities arise.

The power of beliefs and self-awareness in shaping our choices, moving us forward, and being resilient through setbacks.

The beliefs we hold about ourselves, others, and how the world works influence how high we set our goals and also our motivation, persistence, and resilience in pursuing them. It's important to invest the time to really get to know and understand what drives us. Taking a long view of our own careers helps us take risks and puts setbacks in context. And setbacks are often temporary if we learn from them.

Relationships and a diverse network—not just technical skills—are the bedrock of long-term career success.

Building mutually supportive relationships is not just about getting the next job. Strong relationships and networks help us be superstars in the jobs we're in. They give us important resources—ideas, information, contacts, opportunities, mentoring, reputation, and support. Giving thought to the structure of our networks and putting effort into building them to be strong both inside and outside our companies and professions pays off.

None of us can do it alone. We need our tribe—role models who inspire us, mentors who advise and push us, sponsors who advocate for us, and colleagues inside and outside of our companies who can provide perspective and encouragement. Building these connections strengthens our resilience and success.

We are happiest and at our most powerful when we are creating our own futures. So be positive, be thoughtful, and forgive yourself when you fall short. Humanity is a superpower available to everyone. Use it.